Peter Cameron

Hymenoptera Orientalis

Or contributions to a knowledge of the Hymenoptera of the Oriental zoological region

Peter Cameron

Hymenoptera Orientalis
Or contributions to a knowledge of the Hymenoptera of the Oriental zoological region

ISBN/EAN: 9783337222246

Printed in Europe, USA, Canada, Australia, Japan

Cover: Foto ©Andreas Hilbeck / pixelio.de

More available books at **www.hansebooks.com**

Volume of the Fourth Series of "MEMOIRS AND PROCEEDINGS OF ...ESTER LITERARY AND PHILOSOPHICAL SOCIETY."

...era Orientalis: or Contributions ...knowledge of the Hymenoptera ...e Oriental Zoological Region. I.

BY

P. CAMERON.

209033

MANCHESTER:
36, GEORGE STREET.
—
1889.

Hymenoptera Orientalis; or Contributions to a knowledge of the Hymenoptera of the Oriental Zoological Region. By P. Cameron. Communicated by John Boyd, Esq.

(*Received March 11th, 1889.*)

PART I.

INTRODUCTION.

Notwithstanding the large number of our countrymen who reside in our East Indian possessions, our knowledge of their insect fauna, even of the Hindostan peninsula, is exceedingly meagre and fragmentary. A good beginning has been made towards the study of the Lepidoptera, but the same can hardly be said of the other orders. As regards the Hymenoptera, excellent work has been done by our distinguished countryman, Mr. A. R. Wallace, more particularly in the Islands; and his labours have been recorded in numerous papers by my late friend, Mr. Frederick Smith, of the British Museum. But, with all that, very much remains to be done before our knowledge of the Oriental Hymenoptera can be fairly stated to be at all adequate. The fact that less than 2,000 species have been recorded from the Oriental region is sufficient evidence of the truth of this statement; and of the need of the attention of Indian residents being directed to such a promising field of entomological study.

My own attention was drawn to the inquiry by Mr. G. A. James Rothney offering to place at my disposal for study the beautiful and extensive collection formed by him during many years' residence in India, chiefly in the Calcutta district. This valuable source of information has been supplemented by Mr. E. C. Cotes, lending me the material in the Calcutta Museum; by a large collection belonging to the Bombay Natural Society, formed by Mr. R. C. Wroughton, District Forest Officer at Poona; and by various small collections, including a small, but very interesting one, made by Mr. George Lewis, in Ceylon.

In order to make this paper as useful as possible, more particularly to Indian residents, I have given:—

(1) A catalogue of all the known species, with their localities, synonyms, habits, &c.

(2) Descriptions of rare or imperfectly known species.

(3) Descriptions of the new species.

(4) A list of all the works and papers relating to the Oriental Hymenoptera, and

(5) Observations on their geographical relations.

Mr. Rothney's collecting was chiefly in the Calcutta district, namely, in the neighbourhood of the City; in Barrackpore, Sittaghui, Samnugga, Ishapue, Serampue, Chandauague, Gusery; at Port Cauumy to the south, Burdwan to the north; Nischindepue to the north-east. Also in Tirhoot, Bengal; Mussourie, North-west Province (in September and October), in Allahabad, North-west Province; and a few species from Dargeeling, Madras, Bombay, and Ceylon.

Mr. Wroughton's collecting is principally from Poona (Dekhan) and Bombay.

SPHEGIDÆ.

AMMOPHILA.

Ammophila, Kirby, *Trans. Linn. Soc.* IV., p. 195.
Psammophila, Dahlbom, *Hym. Ent.* I., p. 16.
Parapsammophila, Taschenberg, *Zeits. f. d. ges. Naturw.* in Halle, XXXIV.

List of species of *Ammophila* known from the Oriental region.

(i.) *Petiole 2-jointed*:

1. ATRIPES, Smith, *Ann. and Mag., Nat. Hist.* IX., 1852, p. 46; *Cat. Hymen. Ins.*, IV., p. 217, 43.
   Hab. India. Common in Calcutta district (*Rothney*). Khandala (Smith), Sumatra, China, Shanghai.

2. BASALIS, Smith, *Cat. Hymen. Ins.* IV., 214, 37.
   Hab. North India, Punjaub.

3. BUDDHA, Cam., *infra*.
   Hab. Calcutta district, not uncommon.

4. DIMIDIATA, Smith, *l. c.* 216, 40.
   Hab. India (Bombay, Madras, N. Bengal).

5. ELEGANS, Smith, *l. c.* 216, 42.
   Hab. North India (Punjaub).

6. FUSCIPENNIS, Smith, *Trans. Linn. Soc. Zool.* VII., p. 187 (1870).
   Hab. Mainpuri, North-west Province.

7. HUMBERTIANA, Saussure, *Reise d. Novara, Hym.* 25.
   Hab. Ceylon.

8. LÆVIGATA, Smith, *l. c.* 215, 39, de Saussure, *Reise d. Novara, Hym.* 23.
   Hab. India (Madras, Guzerat), Barrackpore (*Rothney*), Ceylon (Cutchevilly).

9. LONGIVENTRIS, Saussure, *l. c.*
   Hab. Ceylon.

10. NIGRIPES, Smith, *l. c.* 215, 38.
    Hab. India (Madras), Barrackpore (*Rothney*).

11. PUNCTATA, Smith, 218, 46.
    Hab. Northern India.

12. ORIENTALIS, Cam., *infra*.
    Hab. Barrackpore, Allahabad (*Rothney*).

13. SMITHII (Baly), Smith, *l. c.* 217, 45.
    Hab. India.

14. SUPERCILIOSA, Saussure, *l. c.* 24.
    Hab. Philippines (Manila).

15. TASCHENBERGI, Cam. *Ammophila erythropus*, Taschenberg, *Zeits. f. d. gesammte Naturw.* XXXIV. 434 (*non* Smith).
    Hab. Java.

16. VAGABUNDA, Smith, *l. c.* p. 218, 47.
    Hab. North China, North India, Sumatra.

17. VISCHU, Cam., *infra*.
    Hab. Mussoorie Hills, North-west Province.

    (ii.) *Petiole with one joint (Psammophila).*

18. HIRTICEPS, Cam., *infra*.
    Hab. Gilgit (Mus. Calcutta).

    (iii.) *Tarsal claws with two teeth at the base (Parapsammophila).*

19. VIOLACEIPENNIS, Cam., *infra*.
    Hab. Sambhalpur, Poonah (*Wroughton*).

20. ERYTHROCEPHALA, Fabricius. *Sphex erythrocephala.* *Fab. Ent. Syst.* II., 204, 23.
    Ammophila erythrocephala, St. Fargeau, *Hist. Nat. Ins. Hym.* III., 385, 26.
    Hab. North India (Punjaub), Poona (*Wroughton*).

A. *Mesothorax transversely striolated.* (*Ammophila, sensu str.*)

AMMOPHILA BUDDHA, sp. nov.

Nigra, fusco hirta, petiolo, scapo, femoribus, tibiis tarsisque, rufis, abdomine cæruleo; alis flavo-hyalinis, apice fere fumatis, nervis testaceis. Long. 25 mm.

Antennæ short, thick; the second joint two-and-a-half times the length of the fourth. Head broad, retreating behind the eyes, which are large and almost parallel; covered with a short sparse white down, and sparsely with longish black hairs; front and vertex obliquely aciculated, the former only excavated immediately above the antennæ and without a longitudinal furrow; clypeus sparsely punctured; its apex almost transverse in the middle, the sides somewhat oblique; the centre slightly incised; mandibles obscure reddish towards the centre, the outer side broadly at the base striolated. Thorax covered with a fuscous pubescence; the tubercles and a spot on either side of the median segment silvery. Pro- and mesonotum strongly transversely striolated, the striolations rather widely separated; propleuræ obliquely striolated; meso- and metapleuræ longitudinally rugosely punctured; metanotum transversely rugosely punctured; scutellum longitudinally striolated; mesonotum with a shallow channel in the centre; metanotum not elevated in the centre; a shallow indistinct furrow below the spiracles. Petiole longish; the second joint usually blackish at the base. Coxæ covered with a dense moderately long silvery pile; the trochanters, tibiæ and tarsi, with a shorter and thinner one; hind coxæ coarsely punctured; tarsal spines black; fore calcaria red; hinder black, reddish at base; apex of tarsi black. Second cubital cellule at top a little wider or a little narrower than the space bounded by the recurrent nervures; third cubital cellule a little wider at top than at bottom, the second transverse cubital nervure bent outwardly at the bottom; tegulæ blackish to piceous.

A. humbertiana, Saus. from Java, seems to be the nearest ally of this species, but it has the metanotum "postice oblique in V-formam elevato-strigato," and the trochanters are not black. *A. basalis* is also nearly related to it, but is smaller (15-17 mm.), has the face silvery pilose, densely so on the clypeus; the head smooth, impunctate, wings hyaline, &c. Barrackpore; Allahabad, N. W. Province.

—AMMOPHILA ORIENTALIS, *sp. nov.*

Nigra, argenteo hirta; femoribus, tibiis, tarsis, petiolo, abdominisque segmento 1° fere toto, rufis, alis hyalinis vel fusco-hyalinis, apice fumatis, costa testacea; nervis nigris; abdomine caeruleo. ♀. Long. 17—19 mm.

Similar to the preceding species, but smaller, with the pubescence shorter and sparser, and of a more silvery tint; the wings without such a decided yellowish tinge, and with the nervures blackish; the first abdominal segment is red, except at the apex, and the third antennal joint is shorter, not being twice the length of fourth. Mandibles broadly red at the base, which is striated; clypeus punctured, densely covered with a silvery pubescence; its apex with a broad shallow sinuation; front and vertex shagreened, sparsely and shortly pilose. Antennæ with the base of first joint testaceous, the flagellum covered with a pale pile. Pro- and mesonotum strongly transversely striolated; metanotum more closely and not so strongly; scutellum strongly longitudinally striolated; propleura perpendicularly striolated, meso- and metapleura obliquely rugosely striolated; the raised part of the metanotum shield-shaped. The tubercles and the sides of the middle segment densely silvery pilose. The second joint of the petiole is black above at the base; the apex has a silky pile; the hind coxæ are white with a dense silvery white pubescence; the trochanters are red, blackish towards the base and apex, the anterior broadly black at the base; the tips of four anterior tarsi and the posterior from the base of the second joint blackish; spurs blackish. Alar cellules pretty much as in *A. buddha*. The ocelli do not form a triangle; the anterior not being placed very far in front of the posterior.

The clypeus and tegulæ in some specimens are testaceous; the apex of the second joint of the petiole may be black; the basal joint of the antennæ may be testaceous, and the middle joints may show a tendency towards fuscous coloration. In size there is some variation.

AMMOPHILA NIGRIPES, *Smith*.

A specimen from Barrackpore agrees with Smith's description so far as it goes. It is fully one line longer; the hair on the thorax is longish and tolerably thick; the clypeus is broadly transverse at the apex, the sides being angled; the mesonotum is furrowed in the centre; the legs are thickly pruinose; the second cubital cellule at the top is about one-fourth shorter than the third, and about equal in length to the space bounded by the second recurrent and second transverse cubital nervures; the third cubital cellule is almost equal in length at top and bottom, and the third transverse cubital nervure is sharply elbowed a little below the middle.

AMMOPHILA ATRIPES, *Smith*.

The Barrackpore specimens of this species, as named by Smith, are uniform in coloration—black, the second joint of petiole is red beneath, the first joint black, the other segments steel-blue; the wings more or less fuscous, the nervures black. Face and clypeus densely covered with silvery white pile; apex of clypeus transverse, the sides rounded; vertex and front with scattered punctures, shining. Pro- and mesonotum strongly transversely striolated; metanotum more closely and not so strongly; scutellum and post scutellum longitudinally striolated; pleuræ rugose. The pubescence on the thorax is short and cinereous; the abdomen is thickly pruinose. At the top the second cubital cellule is about one-half the length of the third, and a little more than the space bounded by the second recurrent and second transverse cubital nervures; the third cubital cellule is nearly equal in length at top and bottom; the third transverse cubital nervure is elbowed near the middle. The female agrees in coloration, punctuation, and clothing with the male.

Differs from *A. nigripes* in being longer, in having the hair on the thorax less dense and shorter, the clypeus more rounded at the apex, the mesonotum with the central furrow less distinct, the wings darker, and with black nervures.

Barrackpore—common.

B. *mesonotum punctured.*

AMMOPHILA VISCHU, *sp. nov.*

Nigra, nitida, punctata; apice petioli, abdominisque segmentis 1—2, rufis; alis fuscis. Long. 22—24 mm.

Antennæ stout, microscopically pilose. Face and clypeus covered with a silvery white pubescence; the front and vertex bear long fuscous hair. Clypeus broad, flat, the apex margined, truncated; sparsely punctured. Front depressed; a distinct furrow down the centre; rather strongly punctured; the vertex with the punctures more widely separated. Thorax strongly punctured, the pleuræ and metanotum rugose; scutellum with the punctures larger and closer than on the mesonotum; post-scutellum rugose. Mesonotum with a distinct furrow, which becomes wider towards the apex, where it is nearly filled up by a keel. The pubescence is long and cinereous, long and dense on the pleuræ; sparser above. The tubercles, an oblique stripe on the pleuræ and the middle segment laterally, densely covered with silvery pubescence. Second segment of petiole stout; the extreme base black. Second segment above wider than the space bounded by the first recurrent and first transverse cubital nervures; the third cellule much narrowed at the top, usually there not one-fourth of the length of the bottom. Tegulæ black.

The male has the clypeus produced and rounded at the apex, and is, as well as the face, densely covered with silvery pubescence.

A. punctata, Smith, is apparently closely allied to this species; but no mention is made of the mesonotum being furrowed, and the metanotum is said to have a longitudinal carina in the centre; the collar has "a minute tubercle in the middle," and the wings are hyaline.

Petiole composed of one joint (Psammophila).

———Ammophila hirticeps, *sp. nov*.

Nigra; longe nigra hirta; abdominis segmentis 2—4 rufis; alis fere hyalinis, apice fumatis, nervis nigris. Long. fere 15 mm.

Antennæ stout; pilose; the third joint about one quarter longer than the fourth. Head hardly punctured; covered with long and black hair; the face and clypeus densely covered with silvery pubescence; apex of clypeus broadly rounded, almost sinuated in the middle; ocelli nearly in a triangle; the posterior separated from the eyes by about the length of the third antennal joint; front hardly depressed. Thorax somewhat punctured; the scutellum apparently indistinctly longitudinally striolated; metanotum obliquely striolated, furrowed down the centre, and with a keel in the centre of the furrow. The one-jointed petiole is a little longer than the second segment, and is covered with long black hair, the fifth segment is red at the base. Above the second and third cubital cellules are sub-equal, and the former above is about three-fourths of the space bounded by the recurrent nervures; the third cellule below is about half the length of the second, and is rounded at the apex below; the third transverse cubital nervure bulges outwardly on the lower half, then retreats towards the second cubital nervure, thus making the third cubital cellule wider below than above. Claws reddish.

Owing to the matting of the hair on the head and thorax, I am unable to make out the sculpture of these parts clearly. The species is a true *Psammophila*.

AMMOPHILA ERYTHROCEPHALA, *Fab.*

This large and striking species is a *Parapsammophila*. The head is large; the eyes reach only exactly opposite the level of the hind ocelli, the vertex being much more developed behind them than usual; they are quite parallel, not converging at the bottom as in *A. violaceipennis*; the antennæ issue from nearly opposite their middle, and not so high up as in the latter species; the clypeus does not project in the middle, and is truncated at the apex. The mandibles are very large and projecting, almost as in *Ampulex*. The neuration of the wings is very much as in *Violaceipennis*. Antennæ black, pilose; the 3—4 basal joints red, the third is nearly twice the length of the fourth.

AMMOPHILA VIOLACEIPENNIS, *sp. nov.*

Nigra; scapo antennarum, petiolo pedibusque, rufis; coxis apiceque tarsorum nigris, alis violaceis. ♂ Long. 29 mm.

Head shining, sparsely punctured; the clypeus and face covered with silvery pubescence; the front and vertex with longish, blackish hair; clypeus somewhat projecting; the apex with a distinct margin, a little sinuated; mandibles broadly red in the middle. The antennæ incline to fuscous beneath, especially at the base; the third joint is longer than the first and second joints united, and about one-fourth longer than the fourth. Thorax densely covered with blackish hair; coarsely punctured; the mesonotum rugosely striolated in the middle at the apex; scutellum coarsely rugosely striolated; metanotum coarsely rugosely punctured in the middle, at the sides obliquely striolated; the pleuræ coarsely rugosely striolated. Pygidium broadly rounded, pilose. Second and third cubital cellules above subequal; the transverse cubital cellules elbowed towards the middle, thus making the third cubital cellule wider in the middle than at top or bottom; the first recurrent nervure is received before the middle of the cellule; the second at nearly the length of the third cubital cellule at the bottom from the apex; at the top the second cubital cellule is as wide as the space bounded by the recurrent nervures.

This species belongs to *Parapsammophila*, Taschenberg, which is chiefly distinguished from *Ammophila* and *Psammophila* by the tarsal claws being bidentate at the base.

PELOPOEUS.

Pelopoeus, Latreille, *Hist. Nat. Ins.* XIII.
Chalybion, Dahlbom, *Hym. Ent.* I., p. 21.

Catalogue of the oriental species of *Pelopoeus*:—

1. P. BENIGNUS, Smith, *Proc. Linn. Soc.* II., 101, 1 *nec*
    P. Javanus, *l. c.* Vol. III., 15, note.
    Hab. Borneo, Singapore, Java.

2. P. BENGALENSIS, Dahlbom, *Hym. Eur.* I., 433, 2.
    Hab. India, Philippines, China, Mauritius.

3. P. BILINEATUS, Smith, *Ann. and Mag. Nat. Hist.* IX., 47 (1852).
    Hab. Bombay.

4. P. COROMANDELICUS, St. Fargeau, *Nat. Hist. Ins. Hym.* III., 302, 2.
    P. fuscus, St. Fargeau, *l. c.* 311, 9.
    Hab. Coromandel, Bengal, Central India.

5. P. CURVATUS, Smith, *Trans. Linn. Soc. Zool.* VII., p. 187.
    Hab. Mainpuri, North-west Provinces.

6. P. FERVENS, Smith, *Proc. Linn. Soc.* II., 101, 2.
    Hab. Java, Borneo.

7. P. JAVANUS, St. Fargeau, *Nat. Hist. Ins. Hym.* III., 306, 9.
    Hab. Java, Malacca.

8. P. MADRASPATANUS, Fabricius, *Syst. Piez.* 203, 3.
   Hab. Bengal, Madras.

9. P. PICTUS, Smith, *Cat. Hym. Ins.* IV., 231, 22.
   Hab. India.

10. P. SEPARATUS, Smith, *Ann. and Mag. Nat. Hist.* IX., 47 (1852).
    Hab. Bombay.

11. P. SOLERI, St. Farg., *Nat. Hist. Ins. Hym.* III., 318, 18.
    Hab. India, (Smith). St. Fargeau gives Guadeloupe as the Habitat of this species.

12. P. SPINOLÆ, St. Farg. *l. c.* 307, 4.
    Hab. Bombay, Ceylon.

13. P. SUMATRANUS, Kohl, *Verh. z.-b. Ges., Wien* 1883, p. 375.
    Hab. Sumatra.

14. VIOLACEUS, Fab., (*Sphex*) *Ent. Syst.* II., p. 201, 12; Lep., *Nat. Hist. Hym.* III., p. 321; André, *Species d. Hym.* III., p. 101.
    Pepsis violacea, Fab., *Syst. Piez.* p. 211, 16.
    Chalybion violaceum, Dbm., *Hym. Ent.*, p. 432, 1.
    Pelopoeus flebilis, Lep., *l. c.*, p. 321, 22.
    Hab. Southern and Eastern Europe, "India," Java.

PELOPOEUS BENGALENSIS.

This is an external builder, erecting its nests on rough walls, or corners, on grass, or on leaves. When on a grass stem the mud is continued far up, thus breaking the outline of the cell, which is in consequence not so readily observed. A solitary cell may be built, or over a dozen may be placed side by side, the whole being then covered well over with mud. (Horne, *Trans. Linn. Soc.* VII. p. 163).

PELOPŒUS MADRASPATANUS.

Of this abundant species (commonly called the mud-dauber) an interesting account is given by Horne (*Trans. Linn. Soc.* VII., p. 161—163). In May, June and July the females are found congregating by small puddles near wells, treading the mud into little pellets of about the size of buck-shot, which, when ready, are brought in the mouth of the insect to the place where the nest is to be constructed. This is in the most various situations. In window-sills, in hollows in walls, in locks, in any cavity between the wall and door-frame; in a depression on the floor, anywhere, in fact, inside or near a house. Horne relates how one individual commenced to build in the corner of a door-frame, where it was crushed every time the door was opened. Six times did the industrious creature commence its habitation only to have it crushed every time. It takes about a day to complete a cell; two, or three, or five are built together, the whole being then covered over with a smooth coating of mud, so that it looks like a dab of mud accidentally left on the wall. When the cell is finished it is filled with small spiders to the number of twenty. Spiders are the regular prey of the *Pelopœus*, but Horne has also seen it store small green caterpillars.

In the pupa state it remains from one to six months according to the season.

PELOPŒUS BILINEATUS.

Unlike *P. Madraspatanus*, this form does not frequent houses, but builds on hedges and trees, a favourite position being a fork in the bough of *Lawsonia spinosa*. As a consequence of the more exposed situation chosen for its nests, these are much more solidly built.

Smith thinks that *P. bilineatus* is only a form of *Madraspatanus*.

PELOPOEUS JAVANUS.

Wallace states (*Jour. Linn. Soc. Zool.* XI., p. 296) that this species enters houses where it constructs small earthen cells, which it stores with paralysed spiders as food for its young. According to Maurice Maindron (*Ann. Soc. Ent. Fr.* 1878, p. 390) the largest nests are 7 centimetres long by 5 in breadth; contain five cells and are made of treaded mud, almost black in colour, but covered in parts by a layer of white earth. The largest and external cell is incomplete and is formed of a whiter earth than the others. In form the nests are irregular and arched; and Wallace (*l. c.*) mentions that they may be plastered over with mud in an irregular manner, so that the shape is completely hidden. The cocoon is $\frac{7}{10}$ of an inch in length, and of a delicate brown colour.

P. COROMANDELICUS.

This species has frequently the scutellum and metanotum without the reddish spot. The clypeus is reddish towards the apex, which is incised in the middle. The mesonotum is transversely striated; the scutellum finely longitudinally striated, but not nearly so strongly as the mesonotum; the pronotum is depressed in the middle; the second cubital cellule is not much narrowed above compared to the bottom, and is broad compared to the length; the first recurrent nervure is received a little before the middle.

SPHEX.

Sphex, Fabricius, *Ent. Syst.* II., p. 198.
Chlorion, Latreille, *Hist. Nat. Crust. et Ins.* IV., p. 57 (*partim*).
Pronæus, Latreille, *loc. cit.* IV., p. 56; Saunders, *Trans. Ent. Soc.* III., p. 58.
Priononyx, Dahlbom, *Hym. Ent.* I., p. 28.
Harpactopus, Smith, *Cat. Hym. Ins.* IV., p. 264.

I. *Tarsal claws with a single tooth near the middle.* = Chlorion, pt. Latr., *Hist. Nat. des Crust. et Ins.* III.; *Pronæus*, Saunders, *Trans. Ent. Soc.* III., p. 58 (1841).

1. SPHEX CHRYSIS.
Sphex cærulea, Christ, (*non* Drury) *Naturg. Ins.* p. 308, tab. 30, fig. 6.
Sphex chrysis, Christ, *l.c.*, p. 310, tab. 30, fig. 7; Kohl, *Termés. Füzetek.* IX., p. 173.
Chlorion lobatum, Fab., *Ent. Syst.* II., p. 206, 30; *Syst. Piez.*, p. 217, 1; Dahlbom, *Hym. Eur.* I., p. 24, 1; St. Fargeau, *Nat. Hist. Hym. Ins.* III., p. 330, 3; Smith, *Cat. Hym.* IV., p. 237.
Chlorion azureum, Lep. et Serv., *Encycl. Méth.* X., p. 451, 2; Lep., *Nat. Hist. Hym. Ins.* III., p. 329.

Common in India all over; also in Burmah, Singapore, Ceylon, China (Hong Kong) Penang and South Africa.

2. SPHEX SPLENDIDA.
Chlorion splendidum, Fabricius, *Syst. Piez.*, p. 218, 5; Smith, *Ann. and Mag. Nat. Hist.* VII., p. 32 (1851).
Sphex pulchra, Lep., *Nat. Hist. Hym. Ins.* III., p. 355.
Pronæus Campbelli, Saunders, *Trans. Ent. Soc.* III., p. 58, tab. 5, fig. 1.
Hab. North India, Burmah, Bombay (Mus. Calcutta), Poona (*Wroughton*).

3. SPHEX MELANOSOMA.
Chlorion melanosoma, Smith, *Cat. Hym. Ins.* IV., p. 238; Magretti, *Bull. Ent. Ital.* XI., p. 578.
Hab. Pondicherry; Kassala (Magretti).

4. SPHEX RUGOSA.
Chlorion rugosum, Smith, *Cat. Hym. Ins.* IV., p. 239.
Hab. Sumatra.

II. *Tarsal claws bidentate; second cubital cellule narrowed towards the radial, higher than long.*—*Harpactopus*.

5. SPHEX ÆGYPTIA.

Sphex ægyptia, Lep., *Nat. Hist. Ins. Hym.* III., p. 181;
Kohl, *Termés. Füzetek* IX., p. 181; Taschenberg,
Zeits. f. d. ges. Naturw. XXXIV., p. 412; André,
Species d. Hym. III., p. 147.

Sphex soror, Dahlbom, *Hym. Ent.* I., p. 436.

Sphex grandis, Radosz., *Hor. Ent. Ross.* XII., p. 132, 2.

Harpactopus crudelis, Smith, *Cat. Hym. Ins.* IV., p. 264, i., pl. vi., fig. 4.

Hab. Eastern Europe, Syria, Egypt, Mauritius, Madras.

6. SPHEX NIVOSA.

Harpactopus nivosus, Smith, *Cat. Hym. Ins.* IV., p. 265, 4.

Hab. North India.

III. *Tarsal claws with three teeth—Enodia.*

7. SPHEX ALBISECTA.

Sphex albisecta, Lep., *Nat. Hist. Ins. Hym.* III., p. 358;
Kohl, *Termés. Füzetek*, p. 185; André, *Species d. Hym.* III., p. 130; J. H. Fabre, *Souvenirs Entomologiques* (1879) p. 174.

Sphex albisecta, Lep. et Serv., *Encycl. Méth.* X., p. 462, 2.

Sphex trichargyra, Spinola, *Am. Soc. Ent. Fr.* VII., p. 466, 11.

Enodia albisecta, Dahlbom, *Hym. Ent.* I., p. 28 and 438;
Costa, *Fauna Reg. Napoli* p. 12, Pl. 1, fig. 3.

Hab. South and Eastern Europe; Africa, from Algiers to the Cape. India.

8. SPHEX PUBESCENS.

Sphex fervens, Fab., *Syst. Ent.* I., p. 346 (*nec* Linné).

Pepsis pubescens, Fab., *Ent. Syst.* II., p. 205.

Enodia canescens, Dahlbom, *Hym. Ent.* IV., p. 28.

Enodia fervens, Dahlbom, *l. c.* p. 439.

Parasphex fervens, Smith, *Cat. Hym. Ins.* IV., p. 267.

Sphex pubescens, Kohl, *Termés.* p. 188; André, *Species d. Hym.* III., p. 130.

Hab. Eastern Europe, Algeria, Guinea, Sierra Leone, Gambia, Cape of Good Hope; India, Madras, Tirhoot (*Rothney*), and North Bengal; China.

IV. *Tarsal claws with two teeth.* (*Sphex sensu str.*).

9. SPHEX. APICALIS.
Sphex apicalis, Smith, *Cat. Hym. Ins.* IV., p. 253 (*non* Smith, *l. c.* p. 262).
Hab. Sumatra.

10. SPHEX ARGENTATA.
Sphex. argentifrons, Lep. *Nat. Hist. Ins. Hym.* III., p. 337; Kohl, *Termés Füzetek* IX., p. 196.
Sphex argentata, Fab. *Ent. Syst.* II., p. 196; Dahlbom, *Hym. Ent.* I., p. 25. André, *Species d. Hym.* III., p. 143; Smith, *Jour. Linn. Soc.* (1869), p. 361.
Sphex albifrons, Lep. *Nat. Hist. Ins. Hym.* III., p. 337, ♂.
Sphex metalica, Taschenberg, *Zeits. f. d. ges. Nat., Halle.* XXXIV., p. 414.
Hab. Eastern Europe, North Africa, China, Japan, India (all over), Ceylon, Java, Amboina, Celebes, New Guinea, Aru, Ceram, Morty Island; Africa, from Egypt to Senegal, Sierra Leone, Angola, Gaboon, Guinea.

11. SPHEX AURIFRONS.
Sphex aurifrons, Smith, *Proc. Linn. Soc.* III., p. 1577, 3.
Hab. Java, Celebes, Aru, Africa.

12. SPHEX AURULENTA.
Sphex aurulenta, Fab., *Ent. Syst.*; Kohl, *Termés. Füzetek* IX., p. 194.
Pepsis sericea, Fab., *Syst. Piez.*, p 211.
Sphex sericea, Dahlbom, *Hym. Ent.* I., p. 26, 7; Lep., *Nat. Hist. d. Ins. Hym.* III., 341, 12.
Sphex fabrecii, Dahlbom, *l. c.* p. 27 and 438.
Sphex ferruginea, Lep., *Nat. Hist. Ins. Hym.* III., p. 345, 18.
Sphex lincola, Lep. *l. c.* p. 353, 27, ♂.

Sphex ferox, Smith, *Jour. Linn. Soc.* IV., p. 55.
Sphex Lepeletierii, Saussure, *Reise d. Novara, Hym.* p. 40, 8.
Sphex Godeffroyi, Saussure, *Stett. Ent. Zeit.* XXX., p. 57.
Hab. China, India, very common in Bengal (*Rothney*), Poona (*Wroughton*), Ceylon, Java, Borneo, Sumatra, Celebes, Amboina, Manilla, Malacca, Ternate, Waigion, Bachian, Ceram, Aru, Timor, Floris, Australia, Cape York.

13. SPHEX ERYTHROPODA, Cam., *infra.*
Hab. India (*Mus. Cal.*).

14. SPHEX FLAVO-VISTATA.
Sphex flavo-vistata, Smith, *Cat. Hym. Ins.* IV., p. 253, 56.
Hab. India.

15. SPHEX NIGRIPES.
Sphex nigripes, Smith, *Cat. Hym. Ins.* IV., p. 253, 56; Kohl. *Termés.* IX., p. 197, 32.
Hab. Hong Kong, Java, Kaschmir.

16. SPHEX ROTHNEYI, Cam., *infra.*
Hab. Allahabad; Mussourie Hills.

17. SPHEX RUFIPENNIS.
Sphex rufipennis, Fab., *Ent. Syst.* II., p. 201, 10; Kohl, *Termés Füzek.*, p. 198, 33; André, *Species d. Hym.* III., p. 149; Lep., *Nat. Hist. Ins. Hym.* III., p. 334, 1; Dahlbom, *Hym. Ent.* I., p. 436, 6; Taschenberg, *Zeits. f. d. g. Naturw., Halle,* XXXIV., p. 411.
Pepsis rufipennis, Fab., *Syst. Piez.*, p. 210, 12.
Sphex diabolicus, Smith, *Proc. Linn. Soc.* II., p. 100, 3.
Sphex fulvipennis, Mocsary, Magy. Ak. *Term. Értek.* XIII.
Hab. North Africa, India; not uncommon in Bengal.

18. SPHEX VICINA.
Sphex vicina, Lep., *Nat. Hist. Ins. Hym.* III., 343, 16.
Hab. India.

19. SPHEX ZANTHOPTERA, Cam., *infra*.

Hab. Barrackpore, Mussourie Hills (*Rothney*).

SPHEX SPLENDIDA, Fab.

Rufa, abdomine negro-cæruleo; alis flavo-hyalinis, apice fumatis, nervis rufo-testaceis. Long. 17 mm.

Scape of antennæ on lower side bearing short black, bristly hairs; the second joint curved inwardly on the inner side; the third thin, more than one-half longer than the fourth. Head almost shining, sparsely covered with black hairs; the front and vertex closely punctured; the face and clypeus more shining, imperceptibly punctured; the labrum and clypeus fringed with short black hairs, the latter with two short stumpy teeth on either side of the middle; a thin furrow runs down from the vertex to the ocelli; the central part of the vertex slightly raised, but not forming a distinct field. Mandibles bearing long black hairs; and some stout furrows towards the middle tooth; the apex is black. Palpi reddish. Thorax shining, sparsely covered with short black hair; the pronotum strongly striolated; the top shining, impunctate, and with a wide and deep furrow in the centre. Mesonotum with scutellum very shining, almost glabrous, sparsely and minutely punctured. Median segment striolated, depressed in the centre and with a furrow along the sides above; the apex rounded, semi-perpendicular, and bearing long black hair; the oblique furrow on pleura is wide and deep, and is divided at the top by an oblique raised projecting part. Abdomen shining; sparsely punctured; pygidial area covered with long black hairs. Legs longish; the hinder row of spines on the hind tibiæ black; the others reddish, and there is a tuft of black spiny hair on the apex of the hinder femora. Tarsal spines thick and stout; metatarsal brush short, thick, reddish. There are some stiff black hair on the hinder tarsi before the claws. Second cubital cellule a little wider at the bottom than at the top, which is a little longer than the top of the third cellule, the latter being very much narrowed at the top, the bottom being more than twice the length of the second cellule, and its apex reaches near to the apex of the radial cellule. The first recurrent nervure is received a little beyond the middle of the cellule; the second quite close to the second transverse cubital nervure.

Sphex aurulenta, Fab.

A variable species. The commonest Bengal form is the var. *aurulenta* Fab. = *Fabricii*, Dbm. = *ferruginea*, Lep. = *godeffroyi*, Saussure. The var. *sericea*, Lep. = *Lepeletierii*, Sauss. also occurs; but I have not seen any Indian specimens that could be referred to the var. *sericea* Fab. = *ferox* Smith, a form chiefly distinguishable from var. *Lepeletierii* by the hair on the pleuræ and middle segment being blackish-brown. The ♂ from Bengal is the typical *lineola* Lep. The hair on the head and thorax is hoary white; the wings are hyaline, smoky at the apex; the abdomen black, the base and the segments at the apices above and beneath reddish; the tegulæ and legs are blackish. A ♂ var also is met with; it has the legs red, except at the base and the tarsi: the tegulæ are red; the hair cinereous; and the abdomen may be red from the petiole, or red only at the base as in the typical *lineola*. This does not quite agree with the description of *S. velox*, Smith, which has the hair fulvous.

Sphex erythropoda, *sp. nov.*

Nigra, fusco pubescens; pedibus rufis; basi apiceque tarsorum, nigris; alis flavo-hyalinis, apice fumatis. Long. 15—18 mm.

Antennæ of the usual length; covered with a sericeous pile; the third joint not much shorter than the fourth and fifth united. Head shining, bearing a scattered punctuation; the front and vertex sparsely covered with longish blackish hair; the cheeks, face, and clypeus densely covered with silvery pile and with longish fuscous hair. Eyes slightly converging beneath; the ocelli hardly forming a triangle; a furrow along their side, the furrows meeting into a V-shaped depression, which has a sharp raised projection in its centre. Clypeus broadly rounded, the apex depressed and with a short incision in the centre. Thorax sparsely covered with a fuscous to black pubescence; the pubescence on the middle segment dull fulvous. Pronotum with a distinct and broad depression in its centre; the mesothorax is also slightly depressed in the centre, and the scutellum and post scutellum are distinctly and broadly furrowed. Median segment transversely and regularly striolated; a wide and deep furrow in its centre at the apex, and there is an elongated pear-shaped depression on the upper part. Abdomen shining, with a plumbeous tint; the petiole covered with long black hair, and a little longer than the coxæ; the pygidial area shagreened, and with a few scattered punctures. Legs with the coxæ, trochanters and four apical joints of the tarsi and the spines on the hinder tibiæ, blackish.

In the colour of the body and pubescence this species comes nearest to *S. rufipennis*, but is readily known from it by the reddish legs. It can hardly, I think, be an extreme variety of *S. aurulenta*, from which, apart from the difference in coloration of the head and thorax and their pubescence (comparing the females), it differs in having the pronotum more distinctly raised above and separated from the mesonotum, besides being broadly furrowed in the centre; the mesonotum and scutellums are also broadly furrowed, and the median segment, instead of having three or four raised ridges, is uniformly and regularly striolated.

The amount of black on the tarsi varies, as does also the colour of the spines and wings, the latter in one specimen having the yellow tint very feebly developed. The tegulæ are for the greater part black.

I have seen four females in the Calcutta Museum collection.

Sphex rufipennis, *Fab.*

This species appears to be a common one in India. The colour of the wings varies, the base, especially in the form *diabolicus*, Smith, being more or less blackish, and the yellow tint is something suffused with fuscous.

S. rufipennis has been recorded from South America, but inasmuch as the ♂ genitalia differs considerably from that of the Indian form, it is probable that the American form, notwithstanding its almost identity in coloration, size, &c., really represents a different species, which I have provisionally named *S. erythroptera* (*Biol. Cent. Am. Hym.* II., p. 30). The form of the scutellum varies in being more or less deeply furrowed. The *S. rufipennis*, Kohl (*Termés. Fuzetek*, IX., p. 198), is, as I am informed by Kohl, a different species from *rufipennis*. Fab. =*luteipennis*, Mocsáry, the latter differing from *rufipennis*, Kohl in having the post scutellum bituberculate, the antennæ thinner, and the wings black at the base.

Sphex argentata.

This large species is common all over the Oriental region, extending also into the Australian Islands of the Malay Archipelago. It is stated by Wallace (*Jour. Linn. Soc.*, XI., p. 296) to be common in the sandy streets of Dobbo, in the Aru Islands, and also at flowering shrubs in Celebes.

Sphex Rothneyi, *sp. nov.*

Nigra; capite et thorace dense et longe argenteo pilosis; abdomine pedibusque rufis; coxis, trochanteribus basique femorum, rufis; alis hyalinis, apice fumatis; clypeo inciso. Long. 22—24 mm.

The face is densely covered with long silvery white hair; the front and vertex densely pubescent and covered sparsely with long gray hair; clypeus rounded. The central incision narrow; eyes slightly converging towards the bottom; mandibles reddish; black at base and apex. Antennæ pubescent; the third joint fully one-half longer than the fourth, which is a little longer than the fifth. Thorax densely covered with a silvery pile; the pronotum above, the metathorax and the pleuræ thickly covered with cinereous hair; a thick line of silvery hair along the tegulæ on the mesonotum; finely punctured; the scutellum shining, bearing distinct punctures, and furrowed down the centre. Median segment with some stout transverse furrows, opaque; rounded and narrowed at the apex and nearly as long as the mesothorax. Petiole black, covered with grey hair; and one-half longer than the hind coxæ. Abdomen shining, indistinctly punctured, elongate, sharply punctured at base and apex; the apical segments more distinctly punctured. Legs longish; broadly black at the base; the tibial spines red; the tarsal reddish in part; the calcaria black, red at the extreme apex. The second cubital cellule is oblique, of equal width at top and bottom and receives the recurrent nervure a very little beyond the middle; the third cellule is longer at the bottom than the second, but at the top is less than one-fourth of the length; the recurrent nervure is received before the middle of the cellule.

The ♂ does not differ in coloration or sculpture from the ♀. The tegulae are reddish. The form of the cubital cellules and the position of the recurrent nervures vary.

In form this species approaches closely to *S. pubescens*; but the black legs of that insect distinguish it at once.

SPHEX XANTHOPTERA, *sp. nov.*

Nigra, argenteo sericeo pubescens; facie, pleuris, pronoto metathoraceque, longe cinereo pilosis; alis flavo-hyalinis apice fumatis. Long. 17—18 mm.

Head closely and minutely punctured; the pile close; the hair on the face and clypeus long and thick; clypeus projecting in the middle, not incised; roundly arched in the male, which has the hair golden; the hair on vertex and front longish, sparse and pale. Mandibles reddish in the middle. Thorax finely punctured; the metanotum transversely striated. The pile is close and dense; on the pronotum above; the mesonotum at the sides; and on the metathorax the hair is longish and dense; on the mesopleuræ it is scarcely so thick. Petiole a little longer than the hind coxæ, densely covered with silvery white hair of moderate length; abdomen sericeous, bluish towards the apex. Legs: coxæ densely covered with long silvery hair; the femora and tibiæ sericeous; the latter thickly spinose; the claws armed at the base with two stout longish teeth. The tibiæ with some stout spines. The second cubital cellule is a little longer at the top than at the bottom, and receives the first recurrent nervure at its extreme apex; the third cubital cellule at the top is one half of the space bounded by the first transverse cubital nervure and the second recurrent.

The male differs in having the hair longer and the pile denser; the clypeus more projecting and broadly rounded at the apex; the abdomen is longer.

TRIROGMA.

Trirogma, Westwood, *Trans. Ent. Soc.* III., 223.
1. *Trirogma cærulea*, Westwood, *l. c.*, 225, t. 12, f. 3 ♂;
   Arc. Ent. II., 66, t. 65, f. 4 ♀.
Hab. Barrackpore (*Rothney*), Poona (*Wroughton*), Northern India, Madras.

AMPULICIDÆ.

RHINOPSIS.

Rhinopsis. Westwood, *Arcana Ent.* II., 68.

Rhinopsis is chiefly distinguished from *Ampulex* by the wings having only three cubital cellules, the first and second being confluent, and by the body not being metallic green or blue.

RHINOPSIS RUFICORNIS, *sp. nov.*

Niger, antennis, ore, thorace, petiolo, tarsisque, rufis ; alis hyalinis, fusco bifasciatis ; nervis sordide testaceis. ♀ Long. 10 mm.

Antennæ shorter than the thorax ; the basal joint curved, as long as the third, which is two-thirds longer than the fourth. Head coarsely alutaceous, almost punctured ; the front keeled, but not distinctly, the keel being interrupted at the base and apex ; eyes parallel. The keel on the clypeus projects at the apex into a stout sharp tooth, and there is a shorter and blunter tooth on either side of this. Prothorax a little shorter than the head ; the top part raised, narrowed and separated from the lower, and deeply furrowed in the centre ; the prosternum and extreme base of pronotum black. Mesonotum shorter than the prothorax, parapsidal furrows slightly diverging at the base, and there is an indistinct furrow between them. Meta- longer than the meso-thorax ; the metanotum with a broad, shallow, somewhat oblique, depression on either side ; in the centre (between the depressions) are three keels, the central straight, the lateral converging towards the apex ; but none of them reach the apex of the metanotum. The metapleuræ are smooth, shining, impunctate ; the rest of the metathorax strongly transversely striolated, running in parts into reticulations. The apex is rounded, margined ; a blunt tooth on either side, and the apex roundly and shallowly incised. The apex is almost perpendicular, broadly furrowed in the centre, and covered with a moderately long white pubescence. Pro- and mesonotum coarsely aciculated, sparsely covered with a white pubescence. Petiole smooth, shining, clavate at the apex ; second abdominal segment as long as all the succeeding segments united; the latter above (especially at their junction), as well as the sides of all, covered with a short pale pubescence. Legs covered with a white pubescence, the femora thickened in the middle, the second cubital cellule is narrowed towards the top; the transverse cubital nervures are straight. Wings not much longer than the thorax.

This species is closely related to the European *R. ruficollis*, Cam., but is much larger, the antennæ and tarsi are red, the metanotum is entirely red, the wings are shorter and not so broadly infuscated in the middle, and with the nervures for the greater part testaceous; and the apex of the petiole is much narrower, thinner, and more club-like.

1. AMPULEX COMPRESSA.

Ampulex, Jurine, *Hym.* 134.
Sphex compressa, Fab., *Ent. Syst.* II., 206, 32.
Ampulex compressa, Dahlbom, *Hym. Eur.* I., p. 29; Lep. *Nat. Hist. Ins. Hym.* III., p. 325, 1; Smith, *Proc. Linn. Soc.* (1869) p. 363.
Chlorion compressum, Fab., *Syst. Piez.* p. 219, 7; Westwood, *Trans. Ent. Soc.* III., p. 227.

A common species, generally distributed over the region. It preys on Blattidæ.

2. AMPULEX HOSPES.

Ampulex hospes, Smith, *Cat. Hym. Ins.* IV., p. 272, 12; *Proc. Linn. Soc.* II., p. 981.
Hab. Borneo.

3. AMPULEX SMARAGDINA.

Ampulex smaragdina, Smith, *Proc. Linn. Soc.* II., 19, 3.
Hab. Singapore.

4. AMPULEX INSULARIS.

Ampulex insularis, Smith, *Proc. Linn. Soc.* II., p. 99, 4.
Hab. Borneo, Malacca.

5. *Ampulex* (?) *annulipes*, Motsulsky, *Bull. Mosc.* XXXVI., (1863).
Hab. Ceylon.

WAAGENIA.

Waagenia, Kriechbaumer, *Ztett. Ent. Zeit.* XXXV., 1874, p. 51.

1. WAAGENIA SIKKIMENSIS, Kriechbaumer, *l. c.*
Hab. Sikkim.

LARRIDAE.

The specific discrimination in this family is at the best a work of some difficulty, and the identification of Smith's species is rendered, in many instances, almost impossible from the absence in his descriptions of any details of structure. Pending an opportunity of studying his types I have left over for future study various species of *Notogonia* and allied genera.

PISON.

Pison, Spinola, *Ins. Lig.*, II., 255; Kohl, *Verh. z.-b. Ges. Wien*, 1884, 180.

1. P. (PARAPISON) AGILE.
Parapison agilis, Smith, *Trans. Ent. Soc.*, 1869, 300, 4.
Hab. Ceylon.

1. P. (PARAPISON) ERYTHROPUS, Kohl.
Parapison rufipes, Smith, *Trans. Ent. Soc.*, 1869, 299, 2; *Trans. Zool. Soc.*, VII., 188, pl. XXI., fig. 1a. (*non* Shuck.)
Hab. Mainpuri, North-west Prov. (*Horne*).

2. P. (PARAPISON) OBLITERATUM,
Pisonoides obliteratus, Smith, *Jour. Proc. Linn. Soc. Zool.*, XII., 1857, 104.
Hab. Borneo (*Wallace*).

3. P. PUNCTIFRONS, Shuckard, *Trans. Ent. Soc.* II., 1837, p. 77, 5.
Hab. " India or St. Helena."

4. P. (PISONITUS) RUGOSUM, Smith, *Cat. Hym. Ins.*, IV., 313, 3.
Pisonites rugosus, Smith, *Trans. Zool. Soc.* VII., 188, pl. XXI., fig. 5a. ♀.
Hab. Mainpuri, North-west Province (*Horne*), Calcutta (*Rothney*), Poona (*Wroughton*).

5. P. SUSPICIOSUM.
Pison suspiciosus, Smith, *Jour. Linn. Soc. Zool.* II. (1857), 104.
Hab. Singapore (*Wallace*).

TRYPOXYLON.

Trypoxylon, Latreille, *Préc. Car. Gen. Ins.*; Kohl, *Verh. z.-b. Ges. Wien.* (1884), 190.

1. TRYPOXYLON ACCUMULATOR, Smith, *Trans. Ent. Soc.* (1875), p. 38.
Hab. Barrackpore (*Rothney*).

2. T. BICOLOR, Smith, *Cat. Hym. Ins.* IV., p. 377.
Hab. Singapore, Java.

3. T. BUDDHA, Cam. *infra*.
Hab. Barrackpore (*Rothney*).

4. T. CANALICULATUM, Cam. *infra*.
Hab. Barrackpore, Mussourie Hills.

5. T. COLORATUM, Smith, *Jour. Linn. Soc. Zool.* II., (1857), 106,
Hab. Borneo (*Wallace*).

6. T. GENICULATUM, Cam. *infra*.
Hab. Barrackpore.

7. T. INTRUDENS, Smith, *Trans. Zool. Soc.* VII., 188.
   Hab. Mainpuri, North-west Provinces (*Horne*), Allahabad (*Rothney*), Ceylon (*Lewis*).

8. T. JAVANUM, Taschenberg, *Zeits. f. d. ges. Naturw.* XLV., 378, 13.
   Hab., Java.

9. NIGRICANS, Cam., *infra*.
   Hab., Barrackpore (*Rothney*).

10. T. PETIOLATUM, Smith, *Jour. Linn. Soc. Zool.* 1857, 105.
    Hab. Borneo (*Wallace*).

11. T. PILIATUM, Smith, *Cat. Hym. Ins.* IV., 377.
    Hab., Madras.

13. T. REJECTOR, Smith, *Trans. Zool. Soc.* VII., p. 189.
    Hab., Mainpuri, North-west Provinces.

14. T. TINCTIPENNE, Cam. *infra*.
    Hab. Barrackpore.

TRYPOXYLON REJECTOR.

The habits of this species are but imperfectly known. Horne found the cells, which are formed of arenaceous mud, and appear very delicate and fragile, but from the strength of the cement used are really tenaceously held together. They are attached to straws usually under cover and constructed chiefly in September.

TRYPOXYLON BUDDHA, *sp. nov.*

Nigrum; fusco pilosum; punctatum; fronte fortiter punctata; metanoto transverse striolato; alis hyalinis; Long. 9—5 mm.

Hab. Barrackpore (*Rothney*).

Antennæ subclavate; covered with a close pile; the third and fourth joints subequal. Head fully broader than the thorax; the front shining, almost bare; the clypeus and lower part of cheeks densely covered with silvery hair. Front raised, furrowed down the centre, bearing large, distinct punctures, narrowed before the antennæ into a wedge. The eyes at top are separated by the length of the second and third antennal joints united; ocelli rather widely separated; clypeus with a raised margin, sharply rounded at the apex. Mandibles reddish beyond the base. Thorax shining, covered with long fuscous hair; mesonotum rather strongly punctured; the scutellum and fore part of the mesopleuræ slightly punctured; the hinder part of the latter impunctate. Metanotum strongly transversely striolated, the striæ wide apart; there is a wedge-shaped depression in the centre of the upper part; the depression with a keel down its edges; there are two lateral keels and the posterior part of the metanotum is widely excavated in the centre; this portion having a gradual rounded curved slope. Petiole as long as the mesothorax, clavate; fully one-third longer than the second segment; the latter is a little longer than the third, and both have an elongate fovea on the top at the apex. At the apex the abdomen is sparsely pilose. Femora sparsely haired; tibiæ and tarsi closely pilose; spurs pale testaceous.

Trypoxylon accumulator.

In this species the front is not much raised on either side of the central furrow, which is wide but shallow; the eyes at the top are separated by about the length of the third antennal joint, at the bottom below the antennæ by fully more than the length of the third. The third joint of the antennæ is nearly one-half longer than the fourth. Clypeus broadly carinate, the apex projecting, broadly rounded. Petiole longer than the thorax, rather abruptly dilated towards the apex; the second segment distinctly shorter than the third.

TRYPOXYLON TINCTIPENNIS, *sp. nov.*

Nigrum; abdominis segmentis 2° et 3° rufis; calcaria albis; clypeo et facie dense argenteo pilosis ; thorace longe albo piloso ; alis fere hyalinis, apice late fuscis. ♀. Long. 12 mm.

Hab. Barrackpore.

Antennæ covered with a silvery down, the third joint about one-fourth longer than the fourth ; the fourth and fifth slightly curved on the lower side. Front and vertex opaque, finely punctured. Front ocellus situated in a pit ; the front before it raised on either side into a roundish elevation, the two being separated by a furrow, at the end of which is a fine straight keel, which reaches near to the base of the antennæ. Eyes at the top separated by the length of the third and fourth joints united ; below reaching to the edge of the clypeus ; below the antennæ they are separated by about the length of the second and third joints united ; clypeus slightly concave, the apex scarcely rounded, being straight to near the centre. Palpi testaceous at apex ; mandibles rufous at tips. The pubescence on the front and vertex is fuscous and very short, on the rest of the head long and silvery white, being especially close and thick below the antennæ. Thorax shining, almost impunctate and with a plumbeous tinge ; the mesonotum bears a sparse short pubescence ; the pleuræ and sternum are more densely covered with longer silvery white hair. At the end of the metanotum there is, in the middle, a bell-shaped depression ; the median segment is deeply depressed in the middle, the depression being widest at the base and continuous with that at apex of metanotum ; its sides are striated. Petiole dilated at the apex and nearly as long as the thorax, and considerably longer than the second and third segments united. The second segment is a little shorter than the third. Legs pruinose, the coxæ bearing longish silvery hair ; the femora are sparsely haired.

Trypoxylon canaliculatum, sp. nov.

Nigrum; palpis, trochanteribus, geniculis, calcaria, tarsisque anterioribus, flavo-albis, tibiæ anticis fulvis; alis hyalinis, apice fere fumatis; tegulis rufo-testaceis; apice petioli abdominisque segmentis 2 et 3 rufis. Long. 9—10 mm.

Antennæ covered with a hoary down; the scape testaceous beneath; the flagellum more or less fuscous; the third joint nearly one-half longer than the fourth. Head opaque, closely punctured; the clypeus, face, cheeks, and eye incision covered with short silvery hair, only visible in certain lights. Front slightly raised, furrowed in the centre; a not very distinct keel at the end of the furrow. Clypeus bluntly carinate in the centre; the apex gaping the margin slightly curved before the middle, which is rounded. Eyes at the top separated by fully the length of the third antennal joint; below the antennæ, by hardly the length of the third. Mandibles rufous. Pro- and mesothorax shining, impunctate; the sides and breast covered with longish white hair. Metanotum shining; a wide depression in the centre, the depression becoming gradually widened to near the apex, which is rounded; on either side of this is a narrow furrow, of nearly equal width and converging towards the apex; both are transversely ribbed; metapleuræ finely obliquely punctured. Median segment widely furrowed in the middle, and covered with white hair. Petiole as long as the thorax, broadly dilated at the apex, and tuberculated at the basal fourth; the second segment distinctly shorter than the third; sides of apical segment distinctly margined laterally; indistinctly keeled in the middle. Legs pruinose; the coxæ bearing white hair.

Hab. Barrackpore, Tirhoot, Mussoorie Hills (*Rothney*).

Trypoxylon piliatum.

Several specimens from Barrackpore are probably referrable to this species. The antennæ bear a short white pile, and have the third joint less than one-fourth longer than the fourth. The checks, eye incision, and clypeus are densely covered with silvery pubescence ; the front and vertex are shining, minutely punctured ; and there is in the latter a large depression, rounded behind, triangular in front, with a distinct raised margin ; from the middle (at the angle) a short keel runs to the eye incision ; and from the apex a stout keel runs to the antennæ. At the top the eyes are separated by the length nearly of the second and third joints united. The two hinder ocelli are placed in round depressions, and are separated by a margin ; the front ocellus is placed in the large frontal area. The metanotum is strongly transversely striolated ; at the base in the centre there is a wide furrow, twice longer than broad, surrounded by a broad margin ; and on either side of this is a broad furrow which unites into a broad furrow running down the centre to the apex. The metapleuræ are much more finely and closely striolated. The mesonotum is finely punctured, and is of almost an olive hue. The abdomen is more than twice the length of the head and thorax united. The petiole is nearly twice the length of the second joint. The calcaria are pale.

The peculiar shield-shaped depression separates this species readily from the others.

TRYPOXYLON INTRUDENS.

Smith has named doubtfully some specimens in Mr. Rothney's collection as this species. They have the head rather strongly punctured ; the front furrowed in the centre ; the eyes at the top separated by the length of the third antennal joint ; there is a wide furrow in the centre of the metanotum, with a curved narrower furrow on either side of it, meeting at the central apical furrow. The furrows transversely striolated ; the rest of the metanotum finely punctured. The petiole is more than half the length of the abdomen, and is dilated not far from the base, and clavate at the apex.

On the whole the specimens agree fairly well with Smith's description, except in what he says about the metanotum, which has "a deep central longitudinal impression; a semicircular enclosed space at the base of the metathorax, which is transversely striolated."

T. intrudens was bred from cells constructed by *Parapison rufipes*.

LARRA.

Larra, Fabricius, *Ent. Syst.* II., 220; Kohl, *Verh. z.-b. Ges., Wien* 1884, 233 (*non* Smith, which = *Stigmus*).

Larrada, Smith, *Cat. Hym. Ins.* IV., 274.

Smith included in *Larrada*, at least three genera, namely, *Larra, Notogonia,* and *Liris* ; probably also *Tachysphex*. From his description it is impossible to make out to which of these groups the majority of his species belong, as he does not mention the structural details on which the genera mentioned are grounded. In these circumstances I have been compelled to leave over for future examination, by means of Smith's types, several species of *Notogonia*. At the best the species are exceedingly difficult to discriminate, the points separating the species being usually minute structural details, most of which are not mentioned by Smith at all.

The following is a list of *Larra sensu lat., i.e.,* of those species which cannot, without an examination of the types, be referred to their precise genus, and which may belong to *Larra, Notogonia, Liris,* or *Tachysphex*.

1. LARRA ALECTO, Smith, *Jour. Linn. Soc. Zool.* II., 103, 6.

Hab. Singapore.

2. L. CARBONARIA, Smith, *l. c.* 102, 2.
   Hab. Singapore.

3. L. CONSPICUA, Smith, *Cat. Hym. Ins.* IV., 276, 7.
   Hab. " India."

4. L. EXILIPES, Smith, *Cat. Hym. Ins.* IV., 278.
   Hab. Northern India.

5. L. EXTENSA, Walker, *Ann. Mag. Nat. Hist.* (3) V., 504.
   Hab. Ceylon.

6. L. LABORIOSA, Smith, *Cat. Hym. Ins.* IV., 278, 12.
   Hab. Philippines.

7. L. MAURA, Fab., *Ent. Syst.* II., 212, 55, Smith, *Cat. Hym. Ins.* IV., 277, 9.
   Hab. Tranquebar.

8. L. POLITA, Smith, *Jour. Linn. Soc. Zool.* II., 102, 4.
   Hab. Borneo, Sarawak.

9. L. SYCORAX, Smith, *Jour. Linn. Soc. Zool.* II., 102, 3.
   Hab. Borneo.

10. L. TISIPHONE, Smith, *Jour. Linn. Soc. Zool.* II., 103, 5.
    Hab. Borneo.

11. L. TRISTIS, Smith, *Cat. Hym. Ins.* IV., 277, 10.
    Hab. Borneo.

12. L. VESTITA, Smith, *Am. Mag. Nat. Hist.* XII., 11.
    Hab. North India.

1. LARRA SIMILLIMA.

Smith, *Cat. Hym. Ins.* IV., 275, 5.

The eyes on the top are separated by the length of the second and third antennal joints united; the vertex has a broad curved depression behind the ocelli and along the sides of the eyes, the centre being raised; there is an indistinct longitudinal furrow in the centre behind; the clypeus is margined, broadly transverse in the middle; the front excavated. The antennæ are stout; covered with a whitish pile; the second joint is half the length of the third. The pronotum has a slight incision in the middle behind; obliquely excavated laterally; shining and finely punctured. The meta- is as long as the mesothorax, and is strongly transversely punctured; the puncturing being much stronger than on the mesothorax; the sides of the metanotum are somewhat depressed; the pleuræ becoming narrowed from the top to the bottom. Pygidial area shining, polished, with a few indistinct scattered punctures along the sides and apex; the sides with a raised margin and with a furrow on the inner side of this margin; the apex broadly rounded, almost truncate.

Of *Larra personata*, Sibi, from Celebes, Smith remarks, "This is probably merely a variety of *L. simillima*, wanting the black apex to the abdomen."

Hab. Tirhoot, Bengal (*Rothney*); "Africa" (Smith *l. c.*).

2. LARRA SUMATRANA.

Kohl, *Verh. z.-b. Ges. Wien*, 1888, 354.
Hab. SUMATRA.

3. LARRA FUSCIPENNIS, sp. nov.

Nigra, argenteo pilosa, abdominis dimidio basali rufo, medio nigro, alis fusco-fumatis, basi sub hyalinis.

Long. 12—13 mm.

Hab. Tirhoot (*Rothney*).

Antennæ short, thick, tapering perceptibly towards the apex; the second joint nearly three-fourths of the length of the third, which is fully one-fourth longer than the fourth. Head shining, strongly punctured; the punctures distinctly separated. Ocellar region not raised and separated; a broad, transverse curved furrow behind it; above the front there is a broad margin. Eyes almost parallel; at the top separated by the length of the third and fourth antennal joints. Clypeus not, or hardly, projecting in the middle; the apex broadly projecting, and with a distinct incision in the middle. Thorax half shining, coarsely punctured; the metathorax more closely punctured than the mesothorax, and densely covered with white hair; pleuræ and sternum shining, the punctures widely separated. Pro- and mesothorax closely covered with dull whitish pubescence. Pronotum in the middle projecting into the mesonotum, which is thus incised broadly. Meta- longer than the mesothorax, the apex perpendicular, indistinctly furrowed in the centre. Abdomen as long as the thorax; covered closely with white pubescence (sparsely on the top of the second and third segments), the apex rather acutely pointed. Pygidial area punctured; covered with a soft white pubescence; the sides not keeled, the apex incised. The basal two segments are red, broadly black in the middle; the ventral segments are pale at their junction, Legs covered with soft cinereous pubescence, tibial and tarsal spines white; calcaria black; metatarsal brush pale. The second cubital cellule at the top is half the length of the third and hardly the length of the space bounded by the current nervures.

Hab. Tirhoot, Bengal (*Rothney*).

4. LARRA NIGRIVENTRIS, *sp. nov.*

Nigra, fere nitida, pruinosa, metathorace opaco, striolato, fere longiore quam mesothorace; alis fere flavo-hyalinis; apice fuscis, nervis flavo-testaceis. Long. 12 mm.

Antennæ the length of the thorax, rather stout, covered with a silvery pile; the second joint one-third of the length of the third, which is hardly one-fourth longer than the fourth. Head wider than the thorax; opaque, alutaceous; eyes at the top separated by more than the length of the third antennal joint; vertex depressed, a wide furrow along either side, close to the eyes; a shallow and less distinct furrow in the centre, leading to and from the ocellus round which it bifurcates, becoming wider and more distinct after leaving it; the presence of the hinder ocelli is not indicated, and the anterior is elongated, being longer than broad, and sharply pointed at base and apex. Face, cheeks, and clypeus densely covered with silvery pubescence. The clypeus slightly projects towards the apex, and is indistinctly carinate down the centre; the apex is broadly rounded, almost truncate. Base of mandibles densely covered with short silvery pubescence; the apex is broadly red, thorax opaque, alutaceous, covered with a sericeous short pubescence; pronotum ending in a rounded part in the centre; mesonotum truncated at base; metathorax nearly longer than the mesothorax; not very distinctly striolated, except at the sides and apex; the latter semi-oblique, furrowed in the middle, the sides densely covered with silvery pile. Abdomen pruinose, hardly longer than the thorax, the apex acute; the pygidial area very shining and bearing a few punctures. Radial cellule not reaching to the apex of the third cubital, wide, and very sharply oblique at the apex; the second cubital cellule shorter than the third, and a very little longer than the space bounded by the recurrent nervures. Legs densely silvery sericeous; the spines and spurs black.

Hab. Barrackpore, Tirhoot; Allahabad, N.W. Provinces (*Rothney*), Poona (*Wroughton*). Not uncommon.

NOTOGONIA.

Notogonia, Costa, *Ann. Mus. Zool. Univ. Napoli* IV., 80 and 82; Kohl, *Verh. z.-b. Ges. Wien*, 1884, 249.

Larrada, Smith = *Tachytes*, Dahlbom, St. Fargeau, Saussure, Taschenberg.

This genus apparently contains more species than either *Larra* or *Liris*.

1. NOTOGONIA PULCHRIPENNIS, *sp. nov.*

Nigra, sericea; mandibulis, tegulis, pedibus (coxis trochanteribusque nigris) abdominisque basi late, rufis, alis flavo-hyalinis, apice fumatis. Long. 12 mm.

Antennæ short, moderately thick; the second joint half the length of the third, the third and fourth subequal. Head almost shining, the face densely covered with silvery pubescence; the vertex with a sparser and shorter pubescence, which does not hide the surface; alutaceous. There is a somewhat triangular depression behind the hinder ocelli; a wide and deep furrow runs down from the anterior, and the depressions on either side of it are deep, curved, and broad. Clypeus not much convex, the apex slightly depressed, and broadly rounded. Eyes at the top separated by the length of the second and third joints united. Thorax densely sericeous, alutaceous, the metathorax transversely striolated, coarsely so at the apex; there is a shallow furrow in the centre of the mesonotum, and there is a narrower and deeper furrow on the apex of the metanotum. The pile on the mesonotum inclines to golden; the metathorax bears a longish white pubescence. Abdomen longer and narrower than the thorax, sericeous; the pygidial area rufous; longitudinally punctured; covered with a silvery pubescence; the sides keeled, the apex rounded, and bearing stiff bristles. Legs moderately sericeous; the bristles and calcaria blackish to fuscous; metatarsal brush silvery. The second cubital cellule is one-third the length of the third at the top, and somewhat less than the space bounded by the recurrent nervures.

Hab. Jeypore (*Rothney*).

2. NOTOGONIA JACULATOR.

Smith, *Cat. Hym. Ins.* IV. p. 279.

In this species the eyes at the top are separated by the

length of the fourth antennal joint; there is a longish shallow ∧-shaped depression above the posterior ocelli; the front depressed where the front ocellus is; and from the apex of the depression a short wide furrow runs; there are three wide depressions on the front above the antennæ, the central being furrowed down the middle. The clypeus is almost transverse. The basal joint of the antennæ is longer than the second and third united; the second is about one-third the length of the third, the latter not being much longer than the fourth. The second cubital cellule is about one-fourth shorter than the third, and wider than the space enclosed by the two recurrent nervures. The pygidial area bears a fulvous to cinereous pile; the apex is broadly rounded. The ♂ has the wings and the nervures darker than in the ♀; the pygidial area has a soft, short, pale pubescence.

Hab. Barrackpore, Mussoorie hills (*Rothney*), Poona (*Wroughton*).

3. NOTOGONIA DEPLANATA, Kohl, *Verh. z.-b. Ges. Wien,* 1883, 358.

Hab. Ceylon.

4. NOTOGONIA SUBTESSELATA, Smith. *Cat. Hym. Ins.* IV., 277, 11.

Hab. Barrackpore. Common (*Rothney*), Poona (*Wroughton*), Sumatra, Java.

A common species.

LIRIS.

Liris, Fabricius, *Syst. Piez.* 227; Kohl, *Verh. z.-b. Ges. Wien,* 1884, 254.

This genus contains, so far as is known, but few species. It is readily known from *Notogonia* by the absence of a notch on the lower side of the mandibles. The pygidial area is clothed with short hair and at the end with stiff bristles; the abdominal segments are usually clothed with a sericeous pile, and the fore tibiæ are spined on the outer side.

1. LIRIS HÆMORRHOIDALIS.

Pompilius hæmorrhoidalis, Fab., *Syst. Piez.* 198.
Liris Savignyi, Spinola, *Ann. Soc. Ent. Fr.* VII., p. 476.
Lyrops aureiventris, Guérin, *Icon. regn. anim.* t. III., 440, pl. LXX. f. 9. ♂.
Liris orichalcea, Dahlbom, *Hym. Ent.* I., 135.
Tachytes illudens, St. Fargeau, *Nat. Hist. Ins. Hym.* III., 249, 12.
Larrada hæmorrhoidalis, Smith, *Cat. Hym. Ins.* IV., 280.
Larrada hæmorrhoidalis, Kohl, *Verh. z.-b. Ges. Wien*, 1884, 256.

A widely distributed species, being found in the Mediterranean region, Syria, Egypt, Senegal, Gambia, Sierra Leone; Punjaub, Poona (*Wroughton*). Smith (*l.c.*) records the species from the Punjaub, but he omits it from his general Catalogue of Indian species (*Trans. Linn. Soc.* 1869).

2. LIRIS AURATUS.

Sphex aurata, Fab., *Ent. Syst.* II., 213, 64.
Liris aurata, Fab., *Syst. Piez.*, p. 228, 3. Kohl, *Verh. z.-b. Ges. Wien*, 1884, 241.
Larrada aurulenta, Smith, *Cat. Hym. Ins.* IV., 276, 6, pl. VII. fig. 5.
Tachytes opulenta, St. Fargeau, *Nat. Hist. Ins. Hym.* III., 246, 7.

Widely distributed. India (common in Calcutta district); Borneo, Sumatra, Java, Bachian, Celebes, China, Japan, Cape of Good Hope, and Gambia.

3. LIRIS NIGRIPENNIS, *sp. nov.*

Nigra, nitida, punctata; facie clypeoque argenteo pilosis; area pygidialis aurea hirsuta; alis fusco-violaceis. Long. ♀ 18; ♂ 15 mm.

Antennæ stout, as long as the thorax. The basal joint keeled on lower side; as long as the second and third joints united; the second joint one-third the length of the third, which is longer than the fourth. Head as wide as the thorax; almost opaque, closely punctured; eyes at the top separated by the length of the fourth antennal joint. A triangular depression above the ocelli, the vertex above this being indistinctly furrowed; there is a wide depression on either side of the ocelli close to the eye; and the space between the upper and lower ocelli is widely furrowed in the middle, the furrow being continued beyond the lower ocellus. The front above the antennæ is widely furrowed along the sides of the eyes, and down the centre. Clypeus distinctly margined at the apex, slightly waved towards the centre. Mandibles black; somewhat hollowed and finely rugose at the base; the apex piceous. Thorax finely punctured; the mesonotum shining, the pleuræ opaque; metanotum also opaque, finely rugose. The pronotum is brought to a point in the middle, and its edge bears a covering of white pubescence; the mesonotum is a little depressed in the centre towards the base; the mesopleural furrow is almost complete; the meta- is shorter than the mesothorax; its apex is semiperpendicular and transversely striolated. Abdomen shorter than the thorax; shining; the segments edged with a pale short silky pile; the pygidial area densely covered with a stiff depressed—golden at the apex, fuscous at the base—pile; and its apex bears stiff golden spines; its surface also bearing stiff blackish bristles. At the top the second cubital cellule is one fourth of the length of the third; the recurrent nervures are almost united, and are received a little before the middle of the cellule. The wings are pale across the cubital cellules. The spines, etc., on the legs are black; the metatarsal brush and the brush on the inner spur dull fulvous.

The ♂ has the hair on the face and clypeus with a more

golden hue; the second cubital cellule is longer in comparison with the third; the recurrent nervures are more widely separated; the pygidial area is less strongly pilose, and wants the bristles on the surface and apex, being also shorter, broader, and with the apex incised.

Hab. Bangalore (*Mus. Cal.*), Poona (*Wroughton*).

PIAGETIA.

PIAGETIA, Ritzema, *Ent. M. Mag.*, IX, 120; Kohl, *Verh. z.-b. Ges. Wien*, 1884, p.

1. *Piagetia Ritsemæ*, Ritzema, *Ent. M. Mag.* IX., p. 120. *Hab.* Sourubuya, Java.

2. PIAGETI RUFICORNIS, *sp. nov.*

Nigra, antennis, ore, clypeo, prothorace, metathorace (medio metanoti nigro) petiolo pedibusque, rufis; alis hyalinis, fascia substigmatali fusca; nervis testaceis. ♀. Long. 9 mm.

Antennæ rather slender, almost bare. The second joint half the length of the fourth, which is shorter than the third. Head wider than the thorax, opaque, finely granular; a furrow runs down from the ocellus to the base of the antennæ, and there is a wider curved furrow on either side of the front; clypeus broadly keeled (the keel narrowed at base), densely covered with a silvery pubescence, the apex with an incision in the middle. Eyes at the top separated by the length of the third antennal joint. Mandibles black at the apical half. Thorax finely aciculated, covered with a close silvery pile; the metanotum finely rugose, with a shallow depression in the centre having a fine keel in the middle. The mesopleuræ and sternum are entirely black; the mesopleural suture rather indistinct; the mesonotum is broadly rufous on either side at the base. Pygidial area almost bare, and marked all over with large punctures. The second cubital cellule at the top is longer than the third; the recurrent nervures are received not far from the base of the cellule, and are almost united. There is a short black line on the top of the middle femora; the posterior femora are entirely lined with black above; the hinder tibiæ are infuscated behind; the coxæ black at the base; the femoral spine is a mere thickening as in *P. ritsemæ*.

May be known from *P. ritsemæ* by there being only a fascia in the wings below the stigma, the entire apex not being infuscated; by the antennal being entirely red; the mesothorax black, &c.; from *P. fasciatiipennis* it differs in being larger; in having the antennæ entirely red; in having the mesonotum broadly red in front; in the mesopleuræ not being entirely black, it being red at base and apex and under the wings; in the metanotum being only black in the middle, the apex too being red; in the second abdominal segment being red at the base; the pygidial area is entirely red and much more strongly punctured; the metathorax can hardly be said to be transversely striated; the wings are not so clearly hyaline, having a fuscous tinge, especially behind the stigma, and the cloud is much more distinct and wider. There is of course, also, the difference in the form of the clypeus and of the femoral spine, but these are doubtless sexual differences which cannot be compared in the absence of the ♂ of *ruficornis* and the ♀ of *fasciatiipennis*.

Hab. Poona (*Wroughton*).

3. *P. fasciatiipennis.* Cameron, *Mem. Lit. and Phil. Soc., Man.* II. (4) 16.

Hab. Ceylon.

TACHYTES.

Tachytes, Panzer, *Krit. Revis.* II., 129; Kohl, *Verh. z.-b. Ges. Wien*, 1884, 327.

Like *Larra* this has been split up into three genera, and the same difficulty is experienced in elucidating Smith's species.

The following are the species which cannot be referred to their proper genus.

1. TACHYTES AURIFEX, Smith, *Jour. Linn. Soc.* II., 101.
   Hab. Borneo.
2. T. FERVIDUS, Smith, *Cat. Hym. Ins.* IV., 298, 11.
   Hab. "India."
3. T. NOVARÆ, Saussure, *Novara Reise, Hym.*, 69.
   Hab. Nicobar Island.

⁻1. TACHYTES ERYTHROPODA, *sp. nov.*

Niger, nitidus, argenteo pubescens; mandibulis, pedibus (coxis nigris) abdominisque segmentis 1—3, *rufo-testaceis; alis hyalinis, apice fere fumatis.* ♀. Long. 8 mm.

Hab. Mussooric hills (*Rothney*).

Head broader than the thorax, shining, sparsely punctured; the vertex sparsely, the checks and clypeus densely covered with long silvery hair. Antennæ short, thick, microscopically pilose; the second joint nearly half the length of the third, which is a little longer than the fourth. Eyes but slightly converging at the top; separated there by the length of the first, second, and third joints united. Ocellar area longer than broad, surrounded by a furrow, and furrowed down the middle; and a furrow winds down from the front ocellus. Lateral prominences indistinct; the clypeus slightly projecting in the middle; the apex in the middle gaping, roundly incised. Thorax shining, impunctate; the pronotum punctured; the metanotum irregularly transversely striolated and covered with long, silvery white hairs. Abdomen longer than the thorax and narrower than it, shining, covered with silvery white pubescence, except on the basal segments in the centre; pygidial area covered closely with stiff fulvous, mixed with white, bristles; the sides keeled; the apex rounded; beneath it is punctured. Femora slightly, tibiæ and tarsi densely covered with white pubescence; tibial and tarsal spines whitish; calcaria rufous; claws for the greater part black. Second cubital cellule about one-fourth longer than the third at the top, and one-half longer than the space bounded by the recurrent nervures.

2. TACHYTES MONETARIUS.

Tachytes monetarius, Smith, *Cat. Hym. Inst.* IV., 298.

The largest and handsomest of the Indian species, and readily known by the abdomen being covered all over with silky golden pubescence. The antennæ have the third joint longer than the fourth, and four times the length of the second. Front and vertex opaque, closely and finely rugosely punctured; eyes at top separated by a little more than the length of the third antennal joint. Clypeus rounded at the apex. Thorax opaque, closely roughly punctured; the medial segment much more strongly than the mesonotum and finely and closely transversely striated at the apex. Second cubital cellule at the top nearly one-fourth shorter than the third; the first recurrent nervure is received about the length of the second cubital cellule from the transverse cubital nervure; the second is received a little beyond the middle of the cellule.

The ♂ has the antennæ stouter; the third joint is distinctly longer than the fourth.

Common, Barrackpore; Mussooric hills (*Rothney*), Poona (*Wroughton*).

3. TACHYTES MODESTUS.

Tachytes modestus, Smith, *Cat. Hym. Ins.* IV., 299.
Saussure, *Hym. Novara Reise*, 72.

This is a larger and stouter insect than *T. ornatipes*; the legs are red, except the coxæ, trochanters and base of femora, the abdomen is shorter, thicker, and more ovate, that of *T. ornatipes* being elongate and narrow; the wings have a more decided yellow tint, and the nervures are more decidedly yellow or rather of a ferruginous colour, but in this respect the wings vary.

Common. Mussoorie hills (*Rothney*). Shanghai (*Saussure*).

4. TACHYTES ORNATIPES, *sp. nov.*

Niger, geniculis, tibiis tarsisque anterioribus, rufo-testaceis; alis fere flavo-hyalinis, nervis testaceis; clypeo, facie thoraceque longe fulvo-hirtis. Long. 12 mm.

Antennæ stout; the third joint hardly longer than the fourth, and three times the length of the second. The hair on the face and clypeus is long and dense, the front and vertex sparsely haired, opaque and sparsely punctured on the vertex, which is depressed and furrowed in the centre. Eyes at top separated by the length of the third antennal joint. Mandibles reddish at the basal half; punctured and covered with silvery-golden hair; palpi reddish testaceous. Thorax opaque. Clypeus punctured; the margin depressed, incised in the middle, the scutellum distinctly punctured; the hair moderately long and thick; the pronotum above with a fringe of silvery pubescence. Abdomen shining; the segments bordered (except in the centre) with silvery pubescence. Pygidial area densely covered with stiff golden hair; sharply narrowed towards the apex, which is rounded. Ventral surface (especially towards the apex) thickly covered with dark brown pubescence and with some scattered longish hairs. Legs cinereous pilose; the femora with scattered hairs; the anterior tibiæ are entirely testaceous; the middle pair are broadly blackish in the centre; the posterior are black, testaceous at base and apex; the hind tarsi black, more or less testaceous at the apex and at the apex of the two basal joints; the spines pale testaceous; the spurs and claws for the greater part rufo-testaceous. The second and third cubital cellules are subequal at the top; the first recurrent nervure is received at a little more than half the length of the top of the second cubital cellule from the transverse cubital nervure; the second a very little beyond the middle of the cellule.

Hab. Barrackpore (*Rothney*).

5. TACHYTES VIRCHU, sp. nov.

Niger, femoribus postecis rufis; capite thoraceque dense fulvo-hirtis; pedibus dense argenteo pilosis; alis fere hyalinis, nervis fuscis. ♂. Long. 8 mm.

Hab. Mussoorie hills (*Rothney*).

Antennæ with the third joint a little shorter than the fourth, and twice the length of the second. Pubescence on clypeus, and face dense, silvery to fulvous; front and vertex bearing long pale fuscous hair; opaque, alutaceous; the vertex rather deeply depressed in the centre. Clypeus with the apex depressed, rounded and shining; thorax with the hair dense and long, opaque; the scutellum finely punctured; the apex of median segment irregularly transversely striated and deeply furrowed in the middle. Abdomen ovate, shorter than the thorax, shining; the segments at the apex with a dense broad silvery fringe slightly interrupted in the middle, except on the apical segment. Pygidial area not much longer than broad, densely covered with depressed silvery hair; the apex broad, truncated. Ventral surface punctured, rather densely covered with dark brown pubescence. Femora behind densely covered with silvery hair; tibiæ and tarsi still more densely with a silvery pile. Spines pale; calcaria fuscous, testaceous at base and apex; claws reddish. Second cubital cellule fully longer than the third at the top; the first recurrent nervure received at the length of the top of the second cubital cellule from the transverse cubital nervure; the second a little beyond the middle of the cellule.

Hab. Mussoorie hills (*Rothney*).

6. TACHYTES ROTHNEYI, sp. nov.

Niger, dense fulvo-hirtus; abdominis segmentis argenteo fasciatis; tibiis tarsisque dense fulvo-pilosis; alis flavo-hyalinis, apice fere fumatis; tegulis rufis. Long. 16—18 mm.

Head and thorax opaque, finely and closely punctured; the scutellum distinctly and strongly punctured; the metanotum at apex irregularly striated and deeply furrowed in the middle. Face and clypeus densely covered with a longish fulvous pile; the vertex sparsely with longish fuscous hair; the occiput with a silvery pile; the mandibles at base with golden pubescence. Eyes at top separated by the length of the fourth antennal joint. Scape of antennæ densely covered with a silvery pile and with some long fuscous hair; the third joint about one fourth longer than the fourth, and three times the length of the second; the fourth—sixth joints are slightly contracted at base and apex, bulging out broadly in the middle. Clypeus broadly carinate in the middle; the apex rounded, entire, and depressed. Mandibles inclining to red towards the apex. Abdomen longer than the thorax; becoming gradually narrowed towards the apex; the basal segment covered with fulvous pubescence; the other segments broadly fringed with silvery pubescence (but the fringe does not extend quite to the middle) at the apex. Pygidial area densely covered with silvery—inclining to golden—depressed stiff pile; its apex truncated. Ventral segments punctured and covered with blackish hair. Tibiæ and tarsi densely covered with fulvous hair, the femora much more thinly; calcaria and spines rufous. The second cubital cellule at the top is nearly one-fourth shorter than the third but at the bottom is longer than it; the first recurrent nervure is received at one-half the length of the second cubital cellule at the top, the second a little beyond the middle, the distance between the two being a little more than the length of the third cubital cellule at the top.

Tirhoot, Bengal (*Rothney*); Calcutta (*Mus. Cal.*).

7. TACHYTES VICINUS, sp. nov.

Niger, dense cinereo hirtus, abdominis segmentis apice pedibusque argenteo pilosis; facie et clypeo longe dense argenteo pilosis; alis fere flavo-hyalinis; tegulis piceis. ♂. Long. 13 mm.

Scape sparsely covered with long pale hair; flagellum opaque, microscopically pubescent: the third joint is, if anything, shorter than the fourth, and not much more than twice the length of the second. Eyes at the top separated by nearly the length of the second and third antennal joints united. Clypeus equally projecting throughout; the apex rounded, hardly depressed. Vertex opaque, alutaceous; sparsely covered with longish fuscous hair; the front bears also long fuscous hair, and laterally a dense silvery pubescence. The silvery pubescence on the clypeus is long and dense. Clypeus distinctly punctured; mandibles still more distinctly and strongly punctured at the base, and bearing a short silvery pile; at the apex they are piceous. Thorax closely punctured all over; at the apex transversely striated. The hair is long and is especially thick on the metathorax. On the sides of the pronotum, and on the mesonotum in front of the tegulæ is a patch of silvery pubescence. The furrow on the apex of the metanotum is narrow and shallow. Abdomen aciculate; the base with sparse fuscous hair; the segments at the apex banded with silvery pubescence, interrupted on the second and third in the middle. Pygidial area with the silvery pile, dense and very bright; the apex roundly incised. Ventral segments at the apices bearing a dense tuft of longish brownish hair, and strongly punctured. Tibiæ and tarsi densely covered with silvery pile; the femora sparsely haired; the calcaria rufous; the tibial and tarsal spines whitish.

Hab. Tirhoot (*Rothney*).

8. TACHYTES NITIDULUS.

Crabro nitidulus, Fabricius, *Ent. Syst.* II., 294, 6 ; *Syst. Piez.* 309, 7.

Tachytes nitidulus, Smith, *Cat. Hym. Ins.* IV., 298 ; Dahlbom, *Hym. Ent.* I., 470.

Tachytes trigonalis, Saussure, *Hym. Novara Reise*, 72.

Common, Barrackpore (*Rothney*), Java.

9. TACHYTES TARSATUS.

Tachytes tarsatus, Smith, *Cat. Hym. Ins.* 296.

A specimen from Barrackpore, and another from Tirhoot, are probably referrable to this species. The antennæ are covered with a pale microscopic down ; the third joint is a little longer than the fourth, and three times the length of the second. Eyes at the top separated by the length of the third antennal joint. Vertex and front almost shining, finely rugosely punctured. Clypeus punctured, the apex depressed, broadly rounded, entire. Thorax closely punctured all over ; the median segment transversely punctured, the apex transversely striated, deeply furrowed down the centre. Abdomen aciculated, punctured closely and finely towards the apex. Pygidial area elongated, sharply pointed at the apex. Ventral surface shining, sparsely haired, aciculated, the apical segments punctured laterally. Wings yellowish hyaline, the nervures yellowish testaceous ; the second cubital cellule one-fourth longer than the second ; the first recurrent nervure is received about the length of the top of the second cubital cellule from the recurrent nervure ; the second about the same distance beyond it, and before the middle of the cellule. The tarsi are only red at the apex.

T. fervidus, Sm., is the only other known Indian species with red abdomen, but it has the legs reddish.

Hab. Tirhoot (*Rothney*).

10. TACHYTES BASALIS, *sp. nov.*

Niger, dense argenteo pilosus; mandibulis, tegulis, scapo antennarum, abdomine dimidio basali apiceque tarsorum, rufis; alis hyalinis, nervis rufo-testaceis. ♀. Long. 10 mm.

Antennæ stout, densely covered with a whitish pile; the third and fourth joints subequal, and about three times longer than the second. Head almost shining; the cheeks, face, and clypeus densely covered with long silvery hair. A narrow but distinct furrow runs down the vertex to the front ocellus, going through the raised ocellar region, which is shining and impunctate at the sides and behind. Clypeus, broadly projecting, becoming sharply turned inwardly before the extreme apex, which thus does not stand on the same plane as the rest of the clypeus; the apex broadly rounded; eyes at the top, separated by about the length of the second and third joints united. Mandibles black at base and apex; the base densely covered with silvery pubescence; the sides bear some long white hairs. Thorax finely and closely punctured; the metathorax finely rugose; its sides and apex densely covered with long silvery hair; the apical furrow rather narrow. Sides of mesonotum bearing close to the tegulæ a broad band of silvery pubescence. The two portions of prothorax subequal laterally; the sternum projecting in front of the fore coxæ. Pleuræ and head densely covered with longish silvery hair. Abdomen shorter than the thorax, shining, aciculate; the segments edged with a fringe of silvery hair. Venter bearing some long fuscous hair. Pygidial area elongate, sharply rounded at the apex; covered with golden, interspersed with silvery bristles; the sides with a not very distinctly raised margin. The coxæ, trochanters and femora in the lower side densely covered with silvery hair; the tibiæ and tarsi densely covered with silvery pile; tibial and tarsal spines pale white; calcaria rufous; outer row of tibial spines rufous; metatarsal brush pale rufous.

Hab. Mussoorie hills (*Rothney*).

TACHYSPHEX.

Tachysphex, Kohl, *Ber. Ent. Zeit.* XXVII., 166; *Verh. z.-b. Ges. Wien*, 1884, 347. = *Tachytes* Auct.

1. TACHYSPHEX ERYTHROGASTER, *sp. nov.*

Niger; capite et thorace dense argenteo pilosis, basi antennarum, clypeo, pedibus abdomineque, rufis, alis clare hyalinis, tegulis pallide rufis, nervis fuscis. ♀. Long. 13 mm.

Antennæ short, stout ; the third joint somewhat shorter than the fourth. Head finely rugose, but the rugosity hid, except in the centre of vertex, by the dense pubescence ; ocellar region raised, broadly, but not deeply, furrowed in the centre ; eyes at the top separated by the length of the third and fourth antennal joints united. Clypeus with an oblique slope at the apex, which is truncated ; labrum with an incision in the middle ; mandibles red, black at the apex ; the base covered with silvery pubescence. Mesonotum and scutellum punctured ; the sculpture of the rest of thorax hid by the dense covering of hair. The apex of metanotum furrowed, perpendicular ; abdomen longer than the head and thorax united, very finely aciculated ; the segments at the apices bearing a band of silky pile ; pygidial area impunctate, narrowing to a point from the middle to the apex ; the sides not very distinctly margined. The second cubital cellule less than one-fourth shorter than the third, and of the length of the space bounded by the recurrent nervures. Legs sparsely pilose, the spines white, the spurs red, the claws blackish.

Hab. Poona (*Wroughton*).

2. TACHYSPHEX ARGYREA.

Larrada Argvrea, Smith, *Cat. Hym. Ins.* IV.

The eyes at the top are separated by fully half the length of the third antennal joint. The part in which are the ocelli is raised; there is a broad transverse depression behind it; a thin furrow is on the top of the vertex, and a wider one runs down from the ocelli. Clypeus bare, shining, impunctate, pale rufous; the apex margined, projecting in the middle. Antennæ filiform rather than stout, densely covered with a pale pile; the second joint is one-third the length of the third. Pronotum rather depressed, having an oblique slope from the top. Pygidial area shining, impunctate, bare, the sides margined, but not stoutly; the apex rather sharply pointed and truncate. The abdominal segments bear laterally a dense silvery pubescence forming broad bands, which do not reach across.

The quantity of black on the abdomen varies, some specimens having the middle segments only slightly infuscated, while others have broad bands on the third—fifth segments. Smith, it may be added, does not state that the clypeus of *Argyrea* is rufous.

Hab. Mussoorie hills (*Rothney*).

3. TACHYSPHEX BENGALENSIS, sp. nov.

Niger, nitidus, punctatus, metathorace rugoso-reticulato, breviore quam mesothorace; alis clare hyalinis, nervis fere nigris. ♀. Long. 10 mm.

Head as broad as the thorax, the vertex sparsely, the cheeks, face and clypeus thickly covered with silvery hair; rather strongly punctured; the eyes at the top separated by the length of the second and third antennal joints united; ocellar region raised; a ∧-shaped depression behind them, with a short longitudinal furrow leading from it, this furrow being continued through the ocellar region itself. Clypeus punctured; margined, and almost truncated at the apex. Mandibles covered with long silvery hair at the basal half. Antennæ nearly as long as the head and thorax united, covered with a dense greyish pile, the third and fourth joints subequal. Thorax shining, bearing a fuscous to silvery pubescence; the metathorax much more thickly than the mesothorax; strongly (especially the pleuræ) punctured; the scutellum not so strongly as the mesonotum. Metathorax shorter than the mesothorax, broader than long, almost rounded at the apex, coarsely rugose, running into reticulations; the apex strongly, nearly transversely striolated. Abdomen as long as the head and thorax united; shining, obscurely shagreened; the segments edged with silvery bands of pubescence, interrupted in the middle; the apex rather acuminate; pygidial area very shining, margined along the side, sparsely punctured. Femora sparsely, tibiæ and tarsi densely covered with white silvery hair; the spines and claws pale ferruginous; the calcaria blackish, reddish on the lower side. The second cubital cellule is about one-fourth longer than the third, the latter at the top being somewhat longer than the space bounded by the recurrent nervures. The apex of the radial cellule is narrow, not sharply angled on the lower part, but rather rounded, and reaches near to the apex of the third cubital. The appendicular cellule is narrow, but distinct.

Hab. Tirhoot (*Rothney*).

— 4. TACHYSPHEX AURICEPS, *sp. nov.*

Niger, aureo-hirtus; pedibus, abdominisque segmentis 1 et 2 rufis, coxis, trochanteribus basique femorum, nigris, alis flavo-hyalinis. ♀ *et* ♂. Long. 12 mm. ♀, 9 mm. ♂.

Antennæ stout, covered with a short white pile; the third and fourth joints subequal. Head as wide as the thorax; the front, cheeks, face, and clypeus covered with a golden pubescence, the vertex with a much shorter and thinner fulvous to golden pile; finely punctured; the eyes at the top separated by the length of nearly the second and third antennal joints united; the vertex furrowed in the centre, the furrow ending in a short ∧-shaped furrow; ocellar region raised, a wide and shallow furrow in the centre, continued down the front as a narrower and more distinct furrow; clypeus at the apex with a distinct, moderately wide margin, rounded and with some small irregular indentations. Mandibles with a red band towards the apex. Thorax covered with a short golden fulvous pile, much longer and thicker on the sides and metathorax; finely and closely punctured; metanotum irregularly transversely rugose, the apex tranversely striolated. Abdomen longer than the thorax; the segments with a broad interrupted band of white pubescence; aciculate; pygidial area with a raised margin along the sides; the apex sharply rounded, bare. Legs shortly pilose; the tibial spines and spurs red; the claws fuscous towards the apex. Second cubital cellule at top half the length of the third, and less than the length of the space bounded by the recurrent nervures, which are received a little in front, and a little beyond the middle respectively.

The ♂ agrees in coloration with the ♀, but the golden pubescence on the head is closer and thicker, the eyes at the top are separated by slightly more than the length of the fourth antennal joint; the third joint is shorter than the fourth; the metanotum is rugose; the two basal joints of the abdomen are banded with black; the wings want the yellowish hue; the second cubital cellule is longer than the third; the nervures are fuscous; and the first transverse cubital nervure is more sharply angled below the middle.

Hab. Poona (*Wroughton*).

GASTROSERICUS.

Gastrosericus, Spinola, *Ann. Soc. Ent. Fr.* VII., 480; Kohl, *Verh. z.-b. Ges. Wien*, 1884, 408.

A genus of small extent, only three species having been hitherto described.

1. GASTROSERICUS WROUGHTONI, *sp. nov.*

Niger, albo pilosus ; tegulis, abdominis segmentis 1—2 *apiceque tarsorum, rufis ; alis hyalinis.* Long. 11 mm.

Antennæ as long as the thorax, densely covered with a silvery pile ; the third and fourth joints subequal, dilated at the apex ; the second one-third of the length of the third. Head fully wider than the thorax ; the cheeks, face, and clypeus densely covered with a silvery pubescence ; the front and vertex much more sparsely. Eyes at the top separated by fully the length of the second and third joints united ; there is a shallow indistinct furrow in the centre of the vertex ; ocelli surrounded by a deep furrow ; hinder ocelli shining, curved, elongated ; vertex and front coarsely aciculated. Apex of clypeus truncated ; mandibles reddish, black at the apex. Thorax punctured, densely covered with cinereous pubescence ; metanotum finely rugose ; its apex perpendicular, almost truncated, but with the sides rounded. Abdomen longer than the thorax, shining, aciculated, the segments broadly banded with a silvery pubescence ; pygidial area bare, except at the apex, which bears long depressed fulvous hair ; the basal portion with scattered punctures. Legs densely covered with silvery pubescence, especially thick on the tibiæ and tarsi ; the anterior tibiæ and tarsi are for the greater part reddish, as are all the knees and spurs ; the spines are whitish. At the top the cubital cellule is somewhat longer than the space bounded by the recurrent nervures, which are received in the basal fourth of the cellule ; the second transverse cubital cellule is curved to near the top, when it becomes angled and straight.

2. GASTROSERICUS ROTHNEYI, *sp. nov.*

Niger, argenteo pilosus, punctatus ; geniculis lineaque tarsorum, albis ; alis hyalinis, apice fere fumatis ; nervis fuscis ; tegulis albis. Long. 7 mm.

Antennæ with a silvery pile; the third and fourth joints subequal. Head closely punctured; the face, cheeks, and clypeus densely covered with long silvery pubescence; eyes almost parallel, at the top separated by the length of the second, third and fourth joints united. Ocellar region raised, roundish, surrounded by a furrow; hinder ocelli as in *G. Wroughtoni*; a narrow indistinct furrow runs down from the front ocellus. Clypeus with a broad truncated projection in the middle at the apex; the middle keeled. Mandibles reddish, black at the base. Thorax finely and closely punctured; the metanotum finely transversely striated, its apex with an oblique slope and furrowed in the middle. The pleuræ and the edge of the pronotum are densely covered with silvery pubescence; the pubescence being especially long on metapleuræ; the tubercles are white. Abdomen aciculate, the segments broadly edged with cinereous pile; pygidial area densely covered with fulvo-golden stiff pubescence. The legs are pilose: the knees, a broad line on the tibiæ behind, the apex of the tarsi and the greater part of the claws are white. The second recurrent nervure is joined to the first before the latter is united to the cubital; the second transverse cubital nervure is not so sharply elbowed as in the preceding species.

Hab. Barrackpore (*Rothney*).

PALARUS.

Palarus, Latreille, *Hist. Nat. Crust. et Ins.* VII., 336; Kohl, *Verh. z.-b. Ges. Wien*, 1884, 416.

1. *Palarus orientalis*, Kohl, *l. c.*, 422.
(?) *Palarus interruptus*, Dahlbom, *Hym. Ent.* I., 468.
   Hab. Ceylon.
2. *Palarus interruptus*, Dahlbom, *Hym. Ent.* I., 468.
   Hab. "Ind. Or."

ASTATA.

Astatus, Latr., *Precis. des caract. gén. des. Ins.*, p. 114, 14.

Astata, Latr., *Hist. Nat. Gén. et part. des Crust. et Inst.* t. III., p. 336.

Over thirty species of this genus are known from various parts of the world, but more particularly from America. Only two have hitherto been recorded from our region.

1. ASTATA MACULIFRONS, *sp. nov.*

Niger, fronte proparte tegulisque flavis; abdominis segmentis 2—5 rufis; alis fusco-hyalinis. ♂. Long. 9 mm.

Antennæ thickened towards the apex, the scape and second and third joints covered with longish hair; the second joint a little longer than the third, and both are perceptibly thinner than the succeeding joints. Front and vertex strongly punctured, almost rugose; the clypeus almost impunctate; the apex broadly rounded; mandibles rugosely punctured at the base; the apex piceous-red. The yellow mark on the front is broader than long, and is rounded at the sides, and is incised in the middle. Pro- and mesothorax shining, sparsely but distinctly punctured; the pleuræ more strongly punctured than the mesonotum; metathorax opaque, coracious, striolated at extreme base; the central part separated from the sides by a curved deep furrow; there is an indistinct keel down the centre, and the apex is rugosely punctured. Abdomen red, the base and the apical two segments black. The second cubital cellule is about two-thirds of the length of the third, and half the length bounded by the recurrent nervures; the first recurrent nervure is received not far from the base; the second a little before the middle of the cellule. The stigma and the nervures beyond its base are testaceous; the apex of the wing is almost hyaline. The legs are covered with long black hair; the anterior knees, tibiæ, and tarsi in front are sordid testaceous, the posterior tarsi have the apices of the joints testaceous.

Hab. Mussoorie hills (*Rothney*).

2. ASTATA AGILIS.

Smith, *Trans. Ent. Soc.*, 1875. 39.

Nigra, facie pleurisque longe argenteo pilosis; abdominis segmentis 1—3 *rufis; metathorace reticulato; alis hyalinis, apice fumatis; tegulis piceis.* ♀. Long. 9 mm.

Antennæ with a close glistening pile; the third joint a little longer than the fourth. Head shining, the front closely but not strongly punctured; the occiput, cheeks, face, and clypeus covered with long silvery hair; there is a short furrow below the front ocellus; the clypeus is rounded at the apex; the mandibles black, reddish in the middle and on the lower side. Thorax shining; the pro- and base of mesonotum closely punctured, the rest of the latter and the scutellum with scattered punctures; the pleuræ coarsely punctured; metanotum longitudinally reticulated; the metapleuræ strongly obliquely striolated; the apex coarsely rugose. Abdomen aciculate; the pygidial area finely rugose; margined at the sides, sharply pointed at the apex. Second cubital cellule half the length of the third and of the space bounded by the recurrent nervures; the first recurrent nervure is received a little before the middle, the second at a somewhat greater distance beyond the middle of the cellule. Tibiæ thickly spined, the apices of the tarsi fuscous.

Hab. Tirhoot, Nischindepore (*Rothney*), Poona (*Wroughton*).

3. ASTATA ORIENTALIS. Smith, *Cat. Hym. Ins.* IV. p. 310, 14.

"India."

This species appears to be closely allied to the preceding, but it differs in having four carinæ on the mesothorax, the wings are flavo-hyaline, clear at the apex, and with ferruginous nervures.

Hab. Nischindipore (*Rothney*).

4. ASTATA ARGENTEOFACIALIS, *sp. nov.*

Nigra, argenteo hirsuta, subtilisme punctata ; metanato rugoso ; abdomine fusco ; alis hyalinis. ♀. Long. 8 mm.

Antennæ covered with a white microscopic pile ; the third joint perceptibly longer than the fourth. Head opaque, coarsely alutaceous ; the occiput, lower part of front, face, and clypeus densely covered with a silvery pubescence ; clypeus incurved in the middle at the apex ; mandibles piceous-red, black in the middle. Thorax opaque, coarsely aciculated ; the metanotum finely rugose, furrowed down the centre, near to the apex above ; the apex oblique, coarsely rugose ; the pleuræ, the pronotum (except in the centre), the sides of the mesonotum ; the hollow at the side of the scutellum, and the sides of the metanotum densely covered with silvery pubescence. Abdomen shining, very finely aciculate ; the segments lined at their junction with a silvery pile ; the basal and apical segments are more or less blackish. Legs covered with a silvery pile ; the spurs and spines white. The second cubital cellule at the top is half the length of the third, and half the length of the space bounded by the recurrent nervures ; at the bottom it is not much shorter than the third ; the first and second transverse cubital nervures are straight ; the first recurrent nervure is received not far from the base of the cellule, the second at nearly double the distance from the apex.

What is apparently the same species has the first and second abdomial segments clear red, and the others quite black.

Hab. Barrackpore (*Rothney*).

ASTATA NIGRICANS, *sp. nov.*

Nigra, nitida, punctata, longe argenteo hirta ; metanoto striolato ; alis hyalinis, nervis, fuscis. ♂. Long. fere 8 mm.

Antennæ as long as the thorax, microscopically pilose,

the joints dilated slightly at the apex; the third joint slightly longer than the fourth. Head (except the ocellar region) densely covered with long silvery hair, moderately punctured; the apex of clypeus rounded; mandibles piceous beyond the middle; the palpi fuscous. Mesonoto and pleuræ punctured, the latter strongly; the metanotum strongly longitudinally striolated, and irregularly reticulated; the hair on the upper part moderately dense, on the sides long and thick; abdomen of the length of the pro- and mesothorax; shining, aciculated; the sides and ventral surface densely covered with long cinereous hair; the segments broadly dull piceous, red at the apices. Legs densely covered with long cinereous hair; the tarsi piceous-red. Second cubital cellule at the top one fourth of the length of the third, and half the length of the space bounded by the recurrent nervures, which are received on either side of the middle of the cellule. The appendicular cellule is incomplete, the nervure ending not far from the radial cellule; the third transverse cubital nervure is angled and issues a short nervure below the middle; the first is sharply angled below the middle.

Hab. Poona (*Wroughton*).

NOTE.—The reference to *Pelopæus violaceus* (p. 12) should be deleted. I now believe, contrary to the opinion of André, that the European *P. violaceus* is not found in India, and is quite distinct from *P. bengalensis.*—P.C., April 15th, 1889.

[*From the Third Volume of the Fourth Series of* "MEMOIRS AND PROCEEDINGS OF THE MANCHESTER LITERARY AND PHILOSOPHICAL SOCIETY." *Session 1889-90.*]

Hymenoptera Orientalis, or Contributions to a knowledge of the Hymenoptera of the Oriental Zoological Region. II.

BY

P. CAMERON.

MANCHESTER:
36, GEORGE STREET

1890.

Hymenoptera Orientalis, or Contributions to a knowledge of the Hymenoptera of the Oriental Zoological Region. By P. Cameron. Communicated by John Boyd, Esq.

(*Received May 13th, 1890.*)

PART II.

NYSSONIDÆ.

GORYTES.

1. *Eyes not converging beneath.*
   (a) *Anal nervure received distinctly before the origin of the cubital nervure in the posterior wings.*

   1. GORYTES ORNATUS.
      Gorytes ornatus, Smith, *Cat. Hym.* IV., 371, 5,[1] Handlirsch, *Sitz. d. Kais. Akad., Wien*, XCVII., 443, 58.[2]
      Northern India,[1] Burmah.[2]
      Unknown to me.

   (b) *Anal nervure received after the origin of the cubital in the hinder wing.*

   2. GORYTES ORIENTALIS (Plate IX., fig. 14 head, 14ᵃ antenna).
      Gorytes tricolor, Smith, *Trans. Ent. Soc.*, 1875, 40. (Nec. Cresson.)
      Gorytes orientalis, Handlirsch, *Sitz. d. Kais. Akad., Wien*, XCVII., 5, 57, 114.

   A very variable species, especially as regards the amount of black, red, and yellow on the head and thorax. The eyes are almost parallel, of moderate size, reaching a little below the base of the clypeus and with moderately large facets; the antennæ with the ♀ stout, thickened towards the apex; the third joint about one-fourth longer than the fourth;

in the ♂ nearly the same relative length and with joints 10—13 hollowed beneath. There is a longitudinal keel on the mesosternum, and there are two perpendicular and one wider and shorter oblique furrows on the mesopleuræ. The first and second transverse cubital nervures are curved; the second cubital cellule at the top is somewhat less than the space bounded by the two recurrent nervures. In the hind wings the anal nervure is received shortly behind the origin of the cubital nervure. The head and thorax bear scattered punctures, those on the head being finer than those on the thorax. The basal area of the median segment is clearly limited by a crenulated furrow; it is broader than long; has two central keels, united by transverse bars; and there are five stout keels on either side of these. The apex of the median segment is hollowed in the centre, and transversely striolated. The basal area varies from black to red. The petiole is moderately broad at the base, and becomes gradually broader towards the apex. Fore tarsi stoutly spined.

Barrackpore (*Rothney*). Several specimens.

3. GORYTES PICTUS.

Gorytes pictus, Smith, *Cat. Hym. Ins.* IV., 365, 22; Handlirsch, *Sitz. d. Kais. Akad. d. Wissen., Wien*, XCVII., 537.

I suspect that this is only a variety of *Orientalis*; but inasmuch as Smith's description gives no information as to what group *G. pictus* belongs, this can only be decided by an examination of the type.

Hab. Madras.

II. *Eyes converging beneath.*

4. GORYTES AMATORIUS (Pl. IX., Fig. 15 head 15[a] antenna).

Gorytes amatorius, Smith, *Trans. Ent. Soc.*, 1875, 39;[1] Handlirsch, *Sitz. d. Kais. Akad. d. Wiss., Wien*, XCVII., 536.

Antennae subclavate, the flagellum thin at the base; the third joint more than twice the length of the fourth. Eyes large, reaching to the base of the mandibles, distinctly converging there, and coarsely facetted. Ocelli almost forming a triangle; separated from each other by twice the length that the posterior are from the eyes. Clypeus moderately convex, the apex margined, almost transverse. Mesosternum without a keel. Basal area of the median segment smooth, shining, glabrous, and limited by a narrow indistinct furrow. Petiole nodose, clearly defined from the second segment, which, at the extreme base, is equal in breadth to it, but from there it becomes much wider. Abdomen elongate, longer than the head and thorax united. Pygidial area clearly limited, covered with short, depressed, coarse, bristle-like hairs, and apparently longitudinally striolated; the sides margined. The second cubital cellule at the top is not very much shorter than the third; and more than the length bounded by the recurrent nervures; the first transverse cubital nervure is angled, near the middle. Fore tarsi and tibiæ spinose.

Barrackpore[1] (*Rothney*). Bombay (*Wroughton*).

NYSSON.

Of the 64 described species of this widely known genus, only two have hitherto been recorded from our region.

1. NYSSON BASALIS.
Nysson basalis, Smith, *Cat. Hym. Ins.* IV., 355, 11;[1]
Handlirsch, *Sitz. d. Kais. Akad. d. Wissen.*, XCVII., 401.
Hab. India.[1]

2. NYSSON DORIAE.
Nysson Doriæ, Gribodo, *Bull. Soc. Ent. Ital.* XVI. 277;[1]
Handlirsch, *l. c.*
Hab. Sarawak. Borneo.[1]

3. NYSSON RUGOSUS (Pl. IX., Fig. 13 head, 13ᵃ antenna)

Niger, linea pronoti, linea abdominis segmentibus 1—2, *maculaque scutelli, flavis, abdominis basi femoribusque posticis rufis; alis fuscis.* Long. 5—7 mm.

Antennæ stout, covered with a white microscopic pile; the third joint not much longer than the fourth; the last conical at apex, longer than the preceding. Head strongly punctured; densely covered with silvery, inclining to golden on the vertex, pubescence. Cheeks margined; an elongated wedge-shaped projection on the front, immediately above the antennæ; eyes slightly sinuated above; apex of clypeus bi-dentate in the middle. Mandibles yellow at base, rufous at the apex. Thorax rugosely punctured, covered with a whitish pubescence; pronotum slightly raised in the middle above, the centre shining, impunctate, glabrous; the sides, looked at from above, curved. Scutellum much more strongly rugose than the mesonotum; metanotum not very clearly defined from it, rounded and narrowed behind; median segment aciculate, shining, bearing four keels; the lateral angles acute, the spine stout, longer than broad. Mesopleuræ convex, densely pilose, strongly punctured, clearly projecting beyond the metapleuræ. Abdomen sparsely covered with shallow distinctly separated punctures, the apical two segments more strongly than the others; pygidial area longitudinally rugosely punctured. Ventral segments punctured like the dorsal; the basal margined laterally; broadly projecting in the middle; the centre of the projection stoutly bicarinate, and its sides are also keeled. Hypopygium aciculate at base and in the centre; the rest punctured, the apex bluntly bi-denticulate. The basal abdominal segment is entirely red, except a yellow line at the apex; the apex of the second segment is yellow; the third at the base is more or less red; the basal two ventral segments are red. Legs (especially the tibiæ and tarsi) densely covered with whitish pile; lateral edges of the coxæ, the knees and apices of the tibiæ more or less yellowish; the hind femora entirely, and the four anterior more or less underneath, rufous; the hind tibiæ more or less rufous beneath. The second cubital cellule is longly pedunculated; the second recurrent nervure is almost interstitial; tegulæ red.

The tibiæ are not spined, nor the clypeus keeled; the metanotum is not bilobate. In some specimens there is a small yellow mark on the sides of the third and fourth abdominal segments.

Hab. Barrackpore (*Rothney*).

4. NYSSON ERYTHROPODA, *sp. nov.* (Pl. IX., Fig. head 18, 18ª antenna).

Niger, argenteo pilosus, capite et thorace rugoso-punctatis; antennis subtus pedibusque rufis, linea pronoti, tegulis, linea scutelli maculisque abdominis segmentis 1—2 flavis; alis fuscis. Long. 6·5 mm.

Agrees closely with *rugosus*; differs in having the antennæ rufous beneath, the legs entirely red, except the base of the coxæ; the base of the abdomen is not red; the yellow line on the pronotum is continuous; the basal joint of the antennæ is more globular, thicker, and shorter; the second is not much shorter than the third; the eyes are more curved, the ocelli more raised; the mesopleuræ at its hinder edge has a stout keel interrupted in the middle, and prolonged beneath near to the sternal groove; the hypopygium is more sharply convex in the middle; the second cubital cellule is longer, being as long, on the lower side, as the third, whereas in *rugosus* it is clearly shorter.

The clypeus is bi-dentate at the apex; the mandibles and palpi obscure rufous; there are four keels on the median segment; its sides are densely covered with pale pubescence; its spine is blunt; the white pile on the tibiæ is very dense, almost hiding the color.

Hab. Barrackpore.

BEMBICIDÆ.

Stizus.

1. **STIZUS BLANDINUS.**
   Larra blandina, Smith, *Cat. Hym. Ins.* IV., 346.[1]
   Hab. India.[1]

2. **STIZUS CORNUTUS.**
   Larra cornuta, Smith, *Ann. Mag. Nat. Hist.*, 1873, p. 403.[1]
   Hab. Bombay.[1]

3. **STIZUS DELESSERTII.**
   Stizus Delesserti, Guérin, *Icon. Règ. Anim.* III., 439.
   Larra Delesserti, Smith, *Cat. Hym. Ins.* IV., 342.[1]
   Hab. Pondicherry.[1]

4. **STIZUS FASCIATUS.**
   Larra fasciata, Fab., *Ent. Syst. Supp.* 253; *Syst. Piez.*, 221, 13; Klug, *Symb. Phys. Dec.* V. t. 46 f. 14? Smith, *Cat. Hym. Ins.* IV., 342.[1]
   Stizus fasciatus, Dahlbom, *Hym. Eur.* I., 133.
   Hab. Barrackpore, Tranquebar,[1] Northern India,[1] Ethiopia.[1]

5. **STIZUS MELANOXANTHA.**
   Larra melanoxantha, Smith, *Cat. Hym. Ins.* IV., 346.[1]
   Hab. India.[1]

6. **STIZUS MELLEUS.**
   Larra mellea, Smith, *Cat. Hym. Ins.* IV., 346.[1]
   Hab. India.[1]

7. **STIZUS NUBILIPENNIS.**
   Larra nubilipennis, Smith, *Cat. Hym. Ins.* IV., 347.[1]
   Hab. India.[1]

8. **STIZUS PRISMATICUS.**
   Larra prismatica, Smith, *Jour. Linn. Soc.*, 1857, 103, 1.[1]
   Stizus prismaticus., Sichel, *Hym. d. Novara Reise*, 142.[2]
   Hab. Borneo;[1] Sambelong.[2]

9. STIZUS REVERSUS.
   Larra reversa, Smith, *Cat. Hym. Ins.* IV., 349.
   Hab. Sumatra.[1]
10. STIZUS RUFESCENS.
    Larra rufescens, Smith, *Cat. Hym. Ins.* IV., 349.
11. STIZUS ORIENTALIS, Cam., *infra*.
    Hab. Barrackpore.
12. STIZUS VESPIFORMIS.
    Sphex vespiformis, Fab., *Spec. Ins.* I., 447, 23.
    Tiphia vespiformis, Fab., *Mant. Ins.* I., 1781.
    Larra vespiformis, Fab., *Ent. Syst.* II., 220, 1,; *Syst. Piez.*, 219, 1.
    Stizus vespiformis, Dahlbom, *Hym. Eur.* I., 154; St.-Fargeau, *Nat. Hist. Hym. Ins.*, III., 297.
    Hab. Madras, Punjaub, Northern India.

STIZUS ORIENTALIS, *sp. nov.*

Brunneus, flavo-variegatus, facie flava; antennis pedibusque rufis; alis flavo-hyalinis. ♀ Long. 24 mm.

Eyes converging towards the apex; ocelli in a triangle; the anterior in a pit; the posterior separated from each other by almost twice the distance they are from the eyes; a broad furrow runs down from the anterior. Clypeus convex; labrum rounded broadly at the apex. A dark brownish stripe runs through the hinder ocelli to the eyes; the anterior ocellus has a similar brownish spot; and there is an elongated mark over each antennæ. The head is closely and finely rugosely punctured. Mandibles clear yellow at the base, shining, impunctate; palish yellow. Scape punctured, covered with a short silvery pile; the third joint is as long as the fourth and fifth united. Thorax finely rugosely punctured; densely covered with soft woolly white hair; median segment at apex semi-perpendicular, deeply and widely sulcated, and transversely striated. The pronotum above and laterally, and the tubercles are yellowish; the mesonotum in front narrowly and more or less of the mesopleuræ are blackish; the apex of median segment also inclining to blackish. Abdomen minutely punctured; a large ovate macula on the sides of the second segment; the greater part of the third and fourth, and the edges of the fifth, clear yellow; the third and fourth ventral segments broadly clear yellow at the sides. Pygidial area punctured. The second cubital cellule is very narrow at the top, not being much more than half the length of the part bounded by the recurrent nervures, and a little more than the space between the second recurrent and the second transverse cubital nervures; the first recurrent nervure is received a little beyond the middle of the cellule. Legs with the tibiæ and tarsi densely covered with a microscopic white pile and stoutly spinose.

STIZUS RUFESCENS, Smith.

A specimen from Barrackpore is no doubt this species—at least it agrees with Smith's description so far as it goes. It is very closely allied to *S. orientalis*; from which, apart from the difference in the coloration of the body, it may be known by the eyes being wider separated, and not, or hardly, converging towards the clypeus; the ocelli are more widely separated from the eyes, and not so much from each other; the radial and second and third cubital cellules are occupied by a deep black cloud: the second cubital cellule is wider at the top: the first recurrent nervure is received before the middle: the apex of the median segment is not so deeply or so widely sulcated; and the wings have a much more yellowish tinge, the nervures and base of stigma being lighter, almost rufous. The length of my specimen is fully one inch.

STIZUS REVERSUS, *Smith*. (Pl. X. f. 1 antenna ♂.)

This species (it is named by Smith in Mr. Rothney's

collection) is common and widely distributed. It shows some variation in the quantity of yellow on the abdomen, as do also the legs, some examples having the tibiæ and tarsi entirely yellow.

BEMBEX.

Of this well-known genus, only four species are known from our region.

1. BEMBEX LUNATA.

    Fabricius, *Syst. Piez.*, 224, 10; Dahlbom, *Hym. Eur.* I, 492, 33; Smith, *Cat. Hym. Inst.* IV, 328, 44.[1]

    Hab. Tranquebar,[1] Tirhoot (*Rothney*), Bombay (*Wroughton*).

    Seemingly the rarest of the species.

2. BEMBEX TREPANDA.

    Dahlbom, *Hym. Eur.* I., 181; Smith, *Jour. Linn. Soc.*, 1869, 366.[1]

    Hab. Barrackpore (*Rothney*); Bombay (*Wroughton*); Gilgit (*Mus. Cal.*); Ceylon (*Rothney*), Celebes,[1] Gilolo.[1]

3. BEMBEX SULPHURESCENS.

    Dahlbom, *Hym. Eur.* I, 180; Smith, *Jour. Linn. Soc.*, 1869, 328.

    Hab. Barrackpore, Tirhoot, Madras (*Rothney*), Punjaub.[1]

4. BEMBEX MELANCHOLICA.

    Smith, *Cat. Hym. Ins.*, IV., 328.[1]

    Hab. China, Sumatra, Borneo, Singapore, Bachian, Celebes, Aru, Salwatty, Morty Islands.[1]

    Said by Wallace (*Jour. Linn. Soc.*, 1869, 296) to be common in sandy situations all over the Malay Archipelago.

PHILANTHIDÆ.

PHILANTHUS.

I have only seen one Indian species of this genus, but Smith records six.

1. PHILANTHUS ELEGANS.
   Smith, *Ann. Mag. Nat. Hist.* XII., 415.
   Hab. Northern India.

2. PHILANTHUS NOTATULLUS.
   Smith, *Proc. Linn. Soc.* V., 157.[1]
   Hab. Menado.[1]

3. PHILANTHUS PULCHERRIMUS.
   Smith, *Cat. Hym. Ins.* IV., 469, 5.[1]
   Hab. India.[1]

4. PHILANTHUS SULPHUREUS.
   Smith, *l. c.* 469.[1]
   Hab. North India.[1]

5. PHILANTHUS DEPREDATOR.
   Smith, *l. c.* 470.[1]
   Hab. India,[1] Barrackpore (*Rothney*).

6. PHILANTHUS BASALIS.
   Smith, *l. c.* 473.[1]
   Hab. Ceylon.[1]

CERCERIS.

1. CERCERIS ALBOPICTA, Smith, *Ann. Mag. Nat. Hist.* XII, 412.[1]
   Hab. Bombay.[1]

2. CERCERIS BIFASCIATA, Guér., *Icon. Règ. An.* 443, Taf. LXXI. f. 9.[1]
   Hab. Bengal.[1]

3. CERCERIS DENTATA, Cam. *postea.*
   Hab. Barrackpore (*Rothney*), Poona (*Wroughton*).

4. CERCERIS EMORTALIS, Saussure, *Reise Novara, Hym*, 98.[1] (See *C. humbertiana.*)
   Hab. Ceylon.[1]

5. CERCERIS FEROX, Smith, *Cat. Hym. Ins.* IV., 454.[1]
   Hab. Sumatra.[1]

6. CERCERIS FERVENS, Smith, *Ann. Mag. Nat. Hist.* XII. 411.[1] (See *postea*).
   Hab. North India.[1]

7. CERCERIS DISSECTA, Fab.
   PHILANTHUS DISSECTUS, Fab., *Ent. Syst. Supp.* 269.[1]
   Hab. India.[1]

8. CERCERIS FLAVOPICTA, Smith, *Cat. Hym. Ins.* IV., 451.
   Hab. North India, Barrackpore, Tirhoot, Madras (*Rothney*).

9. CERCERIS HILARIS, Smith. *Cat. Hym. Ins.* IV., 452.[1]
   Hab. North India,[1] Madras (*Rothney*).

10. CERCERIS HUMBERTIANA, Saussure, *Hym. d. Novara Reise*, 97.
    Cerceris rufinodis, Smith, *Trans. Ent. Soc.*, 1875, p. 41.
    Cerceris viscosus, Smith, *Trans. Ent. Soc.*,[1] 875, 40.
    Hab. Ceylon, Barrackpore.

11. CERCERIS INSTABILIS, Smith, *Cat. Hym. Ins.* IV., 452 ;[1] Saussure, *Hym. d. Novara Reise*, 92.[2]
    Cerceris velox, Smith, *Trans. Ent. Soc.*, 1875, 41.
    Hab. China; Barrackpore (*Rothney*), Poona (*Wroughton*), Ceylon.[2]

12. CERCERIS INTERSTINCTA, Fabricius.
    Philanthus interstinctus, Fab., *Ent. Syst. Supp.*, 269; *Syst. Piez.*, 306.[1]
    Hab. India.[1]

13. CERCERIS MASTOGASTER, Smith, *Cat. Hym. Ins.* IV., 453.[1]
    Hab. Madras.[1]

14. CERCERIS NOVARÆ, Saussure, *Hym. d. Novara Reise*, 92, Taf. IV., f. 54.[1]
    Hab. Ceylon,[1] Bombay, Barrackpore (*Rothney*), Poona (*Wroughton*).

15. CERCERIS NEBULOSA, Cam. *postea*.
    Hab. North Khasi (*Goodwin-Austin*).
16. CERCERIS ORIENTALIS, Smith, *Cat. Hym. Ins.* IV., 54.
    Hab. Madras,[1] Barrackpore.
17. CERCERIS PENTADONTA, Cam. *postea*.
    Hab. Barrackpore (*Rothney*).
18. CERCERIS PICTIVENTRIS, Dahlbom, *Hym. Eur.* I., 498.[2]
    Cerceris pictiventris, Guerstacker, *Monatsch. Berl. Akad. d. Wiss.*, 509; Peters, *Reis. Mozambique*, V., 474.[1]
    Hab. West Africa,[1] Java.[2]
19. CERCERIS PULCHRA, Cam. *postea*.
    Barrackpore (*Rothney*), common, Poona (*Wroughton*).
20. CERCERIS ROTHNEYI, Cam., *postea*.
    Hab. Barrackpore (*Rothney*).
21. CERCERIS SEPULCRALIS, Smith, *Proc. Linn. Soc.* II., 107.[1]
    Hab. Borneo.[1]
22. CERCERIS SULPHUREA, Cam. *postea*.
    Hab. Bombay (*Rothney*).
23. CERCERIS TRISTIS, Cam. *postea*.
    Hab. Barrackpore, Tirhoot (*Rothney*). Common.
24. CERCERIS TETRADONTA, Cam. *postea*.
    Hab. Poona (*Wroughton*).
25. CERCERIS VIGILANS, Smith, *Cat. Hym. Ins.*, IV., 454.
    Hab. Madras,[1] Barrackpore (*Rothney*). Common; Poona (*Wroughton*).
26. CERCERIS VISCHNU, Cam. *postea*.
    Hab. Barrackpore (*Rothney*), Poona (*Wroughton*).

A. *Clypeus in ♀ not porret.*
Cerceris viligans, Smith (Pl. X., f. 2, *a*, *b*).

The ♂ only is described by Smith. The ♀ has the clypeus convex, ending in a rounded point before the apex, from which point it goes obliquely to the labrum, ending on either side in a blunt tooth, the teeth forming with the apical point a triangle. Eyes almost parallel. Ocelli not forming a triangle, the hinder separated from the eyes by the length of the second and third antennal joints united; and from each other by not very much more than the length of the second joint. The third antennal joint is fully one half longer than the fourth, and not much shorter than the first. There is a short blunt triangular tooth on the hinder edge of the mesosternum. The punctuation of the head is moderate; on the thorax much coarser and rugose; the metanotum is rugose; the triangular area on the median segment is large, transversely striated, and indistinctly furrowed down the middle. The pygidial area is opaque, irregularly punctured; twice broader at the base than at the apex, the contraction taking place beyond the middle; the apex rounded, the sides punctured, the hair fringe long, dense, and fulvous. Hypopygium finely punctured, opaque; the incision triangular, reaching a very little beyond the middle.

In the ♂ the clypeus is equally convex, bluntly keeled in the middle, the apex broadly rounded; the triangular area on the median segment is longitudinally striolate, the striæ wide apart; the pygidial area is of equal width and is punctured irregularly.

In both sexes the petiole is wider than long, sparsely punctured; the sculpture on the other segments is weak and becomes almost obsolete on the fifth segment.

CERCERIS ROTHNEYI, *sp. nov.* (Pl. X., f. 3, *a*, *b*).

Ferruginous, the vertex for the greater part black; the fourth and apical segments of abdomen dark piceous; the scape beneath, the cheeks, clypeus, basal half of mandibles, tegulæ, a line on the pronotum, tubercles, scutellum, metanotum, the base of second abdominal segment, the third segment, except on semi-circular space at the base, the fifth segment, the basal two ventral segments and the penultimate on either side and the legs, yellow; the femora lined with fuscous above. Wings clear hyaline, the apex infuscated, the stigma and nervures fuscous. Body covered closely with whitish pubescence, which is especially thick on the head. The punctuation of moderate intensity; the trigonal area of median segment smooth, impunctate, and there is an impunctate space on the sides of the median segment and on the metapleura. Clypeus moderately convex, the apex transverse, with a short blunt tooth at either end. Hinder ocelli separated from the eyes by the length of the third antennal joint, and by a slightly less distance from each other. Eyes a little diverging beneath, the orbits behind clearly margined. The scutellum is smooth, with a border of widely set apart punctures; the metanotum is impunctate. The furrow down the centre of the trigonal area is shallow; the margin of the area is crenulated. Petiole a little longer than wide. Pygidial area elongate, narrow, rounded at the apex; the base narrower than the apex; the top closely and finely transversely rugose; the hair fringe white; hypopygium not incised. Antennæ blackish above; the flagellum obscure rufous beneath. Second cubital cellule with the peduncle not much shorter than the width of the cellule itself; the first recurrent nervure is received in the basal third; the second a little beyond the apex of the cellule, and almost interstitial. Length 10 mm.

This species comes nearest to *C. instabilis;* but may easily be known from it by the clypeus not being incised; by the punctuation being not so strong, especially on the scutellum and median segment; by the smooth, not rugosely punctured, trigonal area, by the narrower, longer, straighter, pygidial area, which in *instabilis* is wider at the base than at the apex, it bearing also large punctures, and by the shorter second cubital cellule.

CERCERIS INSTABILIS, *Smith*, (Pl. X. f. 4—4 *a*, *b*).

A common and variable species, especially as regards the amount of black on the thorax and abdomen. The ♂ is *C. velox*, Smith.

CERCERIS PULCHRA. (Pl. X. f. 5, *a*, *b*.)

Black, the scape beneath, the cheeks, clypeus, mandibles except at apex, two lines on the pronotum, tegulæ, metanotum, a mark on the base of the second abdominal segment, the third segment except the base in the centre; the fifth segment, the second ventral segment laterally, and the legs, yellow; the petiole and second abdominal segment ferruginous, except the yellow and a black mark on the latter. Clypeus a little gaping at the apex, which is black and incised. Eyes parallel. Hinder ocelli separated from the eyes by near the same distance they are from each other. Body covered with longish pale fulvous pubescence, almost golden on the face; the punctuation is strong and coarse; the scutellum strongly punctured. Trigonal region rugose, and with some stout keels; the central furrow deep. Petiole distinctly longer than broad; pygidial area elongate, gradually narrowed towards the apex, which is almost transverse, but with the edges rounded; the incision in the hypopygium is a little longer than wide, rounded at the base, and becoming wider towards the apex; the top of the pygidial area is irregularly punctured, without a hair fringe, and covered with long pale pubescence. The coxæ are black at the base; the hinder femora are for the greater part black; the hind tibiæ are fuscous on the outer side at the apex. Antennæ rufous; the third joint curved, twice the length of the second and longer than the fourth. The wings are suffused with fuscous; the apex broadly smoky; the second cubital cellule has the peduncle not much shorter than the cellule; the first recurrent nervure is received in the basal fourth; the second a little beyond the second transverse cubital nervure; the nervures fuscous.

The ♂ has the clypeus at the apex broadly rounded; the black on the second abdominal segment is more extended; the fifth segment is black and the sixth yellow; the pygidial area is punctured strongly, narrowed a little at the base, and with the apex transverse. Length 7—8 mm.

This species is represented by numerous examples of both sexes, and it appears to be tolerably constant in coloration. *Vischnu* may be known from it by the black face, by the truncated clypeus, by the absence of yellow on the thorax and on the second abdominal segment; *dentata* by the teeth on the thorax.

CERCERIS VISCHNU. (Pl. X., Fig. 7—7*a*, *b*.)

Black; the scape, the apex of the clypeus, the antennal ridge, a mark on the cheeks, a small mark on the sides of the pronotum, the apex of the third abdominal segment, a mark in the centre of the fifth, the four anterior tibiæ in front, the four anterior tibiæ, the base of the hinder, pale yellow; the femora beneath and the hind coxæ rufous; the petiole and base of second segment ferruginous. Clypeus with the apex projecting, but not sharply and with a slight incision. Ocelli in a curve; the posterior separated from each other by more than the length of the third antennal joint and by its length from each other. The face covered with a silvery pubescence, the rest of the body with pale hair; the punctuation close, rugose. Trigonal area not very clearly defined, rugosely punctured. Petiole a little longer than broad; pygidial area rugose, slightly narrowed towards the apex, which is bluntly rounded; the hair fringe dense, obscure fulvous; the incision in hypopygium a little longer than broad, the base rather sharply pointed. Wings clear hyaline, the apex smoky; the second cubital cellule longer than broad; the recurrent nervure received shortly before the middle. The antennæ have the flagellum pale fulvous beneath; the eyes are parallel.

The ♂ has the clypeus broadly truncate at the apex in the middle, and with the lateral laminæ pale; the clypeus wants the yellow mark; the petiole is black; otherwise as in ♀ except that it is the 6th abdominal segment which is marked with yellow. Length ♀ nearly 9 mm; ♂ 7 mm.

A distinct species.

What appears to be this species has a small yellow mark at the base of the abdomen, the petiole beneath and laterally and the base of the second segment are rufous and there is a small white mark on the base of the latter.

CERCERIS TRISTIS, *sp. nov.* (Pl. X. f. 7, *a*, *b*).

Eyes parallel. Ocelli not forming a triangle; the posterior separated from the eyes by nearly the length of the second and third joints united and from each other by less than the length of the third. Middle lobe of the clypeus convex, gaping at the apex, which is broadly and narrowly incised. Punctuation of head close and rather strong; the face densely covered with silvery hair. Head below the antennæ and the orbits for the length of the scape above the base of the antennæ, and the mandibles, except at the apex, yellow. Extreme apex of clypeus black. Thorax strongly punctured, more coarsely on the pleuræ; the tegulæ, a line on either side of the pronotum, and metanotum, yellow. Trigonal area obscurely aciculated, the sides punctured irregularly; the central furrow crenulated. Petiole longer than broad, distinctly bulging out at the sides; the segments with a moderately strong punctuation; a mark broader than long on the base of the second segment; the apex of the third segment (narrowed in the centre) and a narrower belt on the fifth, also narrowed in the centre, yellow. Pygidial area elongated, narrower at base than at apex, the latter bluntly rounded; the surface irregularly rugose. Incision in hypopygium broad, a little longer than broad, gaping at the apex, the base broadly rounded. Lateral hair fringe sparse, silvery. Legs black; the apex of coxæ and trochanters; the greater part of the femora beneath; and the tibiæ and tarsi, except the apices of the posterior, yellow.

The ♂ has the clypeus slightly convex; the apex projecting in the middle, black and with a short broad, blunt tooth in the middle. The coloration is as in the ♀ except that the fifth segment is entirely black and the sixth nearly entirely yellow. Pygidial area of equal width throughout; depressed before the apex, shining and sparsely punctured. Length 7—9 mm.

May be known from *C. novaræ* by the median segment wanting yellow marks; by there being only a central basal yellow mark on the second abdominal segment; by the legs being broadly black at the base, by the third joint of the flagellum being distinctly longer than the fourth; by the central lobe of clypeus being roundly convex and gaping at the apex, &c.

CERCERIS FERVENS, *Smith*.

It is probable that this will prove to be identical with *C. novaræ*. So far as the description of the ♀ goes the only difference is that in *fervens* there are two lines on the scutellum; but in both the females of *Novaræ* I have examined, there is a minute yellow mark on either side of it. Smith's description of the ♂ differs moreover, it having the sixth segment (not the fifth) yellow marked; and apparently the median segment wants the yellow marks.

Smith, I may add, named, in Mr. Rothney's collection, the ♂ of the species I have called *Vischnu, fervens*; but it does not agree with the description of *fervens*; and my specimen is certainly not the ♂ of *Novaræ*. It is possible that Smith has assigned the wrong ♂ to his *fervens*.

CERCERIS NOVARÆ, *Sauss.*, (Pl. X. f. 8, *a, b*).

Clypeus flat, slightly convex in the middle; the apex projecting a little beyond the lateral pieces; the apex truncated, but with the sides rounded. Eyes a little converging beneath. Ocelli not forming a triangle, the anterior being too much in front of the posterior; the posterior separated from the eyes by a little more than the length of the third antennal joint, and by its length from each other. Front and vertex strongly punctured; the face and clypeus with shallow punctures. The face, oral region, antennal keel and the orbits to the length of the scape above the base of the antennæ; the mandibles, except at the apex, and a spot behind the eye near the top, yellow. The vertex is sparsely pilose. Thorax more strongly punctured than the head; the punctures larger and wider apart, the pleuræ more coarsely punctured than the mesonotum. Trigonal area impunctate, furrowed down the centre, the furrow crenulated. Two marks on the pronotum, a spot on the mesopleura, tegulæ, a small mark on the sides of the scutellum, the metanotum and two large oblong marks on the median segment, yellow. Mesopleural furrow, wide, deep. Apex of median segment gradually rounded above, the apical part rather abrupt, and with two deep, shining, somewhat triangular depressions at the extreme apex. Petiole longer than broad, slightly bulging out at the middle, coarsely punctured. The other segments are also strongly punctured. Pygidial area an elongated oval, narrowed at base and apex, the surface irregularly rugose. Hypopygial incision not reaching to the middle of the segment, longer than broad, rather acutely pointed at the apex. The lateral hair fringe dense, golden-fulvous. There is a broad band on the base of the second segment, the whole of the third segment, except a somewhat semicircular black mark in the middle at the base and a similar amount of black on the fifth, yellow. The legs are yellow, except the apical half of the posterior femora and tibiæ. The scape is yellow; the flagellum fulvous beneath; the third and fourth joints are sub-equal.

The ♂ is similarly coloured, except that the mesopleuræ want the yellow mark and the scutellum is entirely yellow. Length 7—10 mm.

CERCERIS WROUGHTONI, *sp. nov.* (Pl. X. f. 9, *a, b*).

Black, shining, thorax, and head punctured, the abdomen impunctate, smooth; the clypeus, the cheeks to above the base of the antennæ, the antennal ridge, the scape beneath, a large broad line behind the eyes, a line on the pronotum, scutellum, metanotum, two small marks on the middle of the median segment, the petiole, except a broad stripe down the centre, the base of the second and third segments all round, the edge of the fourth, and a large mark in its centre, a similar mark on the fifth; and the basal three ventral segments, for the greater part, clear yellow; the flagellum beneath and a large mark on either side of the median segment rufo-fulvous; legs fulvous, the anterior four in front and the hinder femora in part, pale yellow. Wings fuscous, the apex much darker, the stigma fulvous. Antennæ with the third joint about one quarter longer than the fourth. Clypeus flat, the apex black, roundly and broadly incised; its sides and the cheeks bearing a silvery pubescence. Ocelli forming almost a triangle, the posterior separated from the eyes by more than the length of the third antennal joint; and by about its length from each other. The pubescence on the head is longish and pale; the punctuation moderately strong, and all the punctures deeply separated. Thorax moderately strongly punctured, covered with a pale pubescence; the pronotum above saddle-shaped; the transverse furrow in the mesopleura wide, deep and complete; scutellum sparsely punctured; trigonal area smooth, impunctate, furrowed down the middle; metapleuræ aciculate, the base with a few stout striæ. Petiole wider than long, bulging out in the centre; sparsely and finely punctured; the fifth segment also sparsely punctured; the others obscurely aciculate. Pygidial area large, rounded at apex, not narrowed at the base; transversely striolated; incision in hypopygium not reaching to the middle, a little longer than wide, rounded at the base. Hinder tibiæ very stoutly and closely spined. Second cubital cellule longer than broad, rounded above, above shortly pedunculated; the recurrent nervure received shortly before the middle.

The ♂ has the face from the top of the antennal ridge entirely yellow; the clypeus flattish; the apex rounded, almost transverse; the lateral plates incised; there is a small yellow mark below the tubercles; there are two large yellow marks on the median segments; the fourth segment is almost entirely yellow; the fifth has a narrow yellow border on the apex, the sixth is almost entirely yellow; the sides of the apical segment are yellow: the legs are clear yellow; the hinder femora are marked with black, the hind tibiæ are broadly black at the apex and the hind tarsi are fuscous; otherwise coloured as in ♀. Pygidial area with large punctures, the apex almost transverse. Length 11 mm.

Easily recognised by the smooth impunctate abdomen.

CERCERIS PICTIVENTRIS, *Dbm.*

This species is no doubt identical with *C. novaræ*, at least that species agrees with the rather laconic description given by Dahlbom—Abdomen nigrum, flavo-fulvo—aut albo-fasciatum . . . petiolus nigrum . . . segmenta ventralia plurima flavo-fasciata aut maculata—Corpus subparvum . . . abdomen fasciis citrinis pleurumque 3 raro pluribus; ♀ segmento ventrali valvutæ proximo ad marginem apicalem intergerrimo.

Schletterer (*Zool. Jahrb.* II., p. 499) quotes doubtfully *C. pictiventris* of Gerstacker (*Monatsb. Berl. Akad. Wiss.*, 509) as a synonym. Gerstacker's species was from West Africa Dahlbom's from Java.

CERCERIS DENTATA, *sp. nov*. (Pl. X., f. 10, *a, b*.)

Black; the clypeus, cheeks, antennal ridge, mandibles except at the apex, a spot behind the eyes, an interrupted line on the pronotum, two marks on the scutellum, two marks on the mesopleuræ, metanotum, two elongated marks on the median segment, two small marks on the petiole, a line on the apex of the second and third abdominal segments, the fifth, except at the base, the femora beneath and the tibiæ and tarsi, clear yellow; the petiole, coxæ, and femora, ferruginous; wings hyaline; very slightly infuscated at the apex. Clypeus depressed, the apex transverse. Eyes a little diverging; ocelli not forming a triangle, the posterior separated from the eyes by the length of the third antennal joint and from each other by the length of the fourth. Body covered with a white pubescence, silvery on the face; the punctuation moderately strong. The mesosternum in the middle projects into two stout teeth; the posterior being the larger. Trigonal area elongated, reaching nearly to the apex of the segment, much longer than broad, smooth, impunctate, a fovea on either side of the furrow. Scutellum almost impunctate. Mesopleuræ rugosely punctured, convex, bulging out above; metapleuræ finely and closely rugose at the base, the apex with large punctures. Petiole hardly longer than broad. Pygidial area narrowed at base and apex, bulging out gradually in the middle, the apex above black, the rest dull rufous, and punctured; the hair fringe dense, obscure white. Ventral surface with second and third segments banded with pale yellow; covered with long white hair; the incision in apical segment reaching near to the middle, rounded at base. Wings shorter than usual; the second cubital cellule twice longer than wide; the recurrent nervure received in the basal third. Length 8—9 mm.

The spines on the mesosternum separate readily this distinct species.

CERCERIS SULPHUREA, *sp. nov.* (Pl. X., f. 11, *a*, *b*).

Sulphureous; the ocellar region and three lines on the mesonotum, sordid rufous; wings clear hyaline, the apex from a little before the end of the radial cellule fuscous, the nervures fuscous; the flagellum of the antennæ pale rufous. Clypeus convex, broadly rounded at the apex. Eyes slightly diverging; posterior ocelli separated from each other by the length of the third and from the eyes by the length of the first antennal joint. The third antennal joint only a little longer than the fourth. Body covered with long white hair; that on the face silvery. Punctuation moderately strong. Trigonal area punctured. Petiole nodose, longer than broad. Pygidial area broad, but longer than broad, punctured, the apex bluntly rounded. Second cubital cellule not much longer than broad; the first recurrent nervure almost interstitial. ♂. Length 8 mm.

CERCERIS TETRADONTA, *sp. nov.* (Pl. X. f. 12, *a*, *b*).

Black, closely and rather strongly punctured, the clypeus, the cheeks broadly from near to the top of the antennal ridge, the antennal ridge; scape beneath, the base of the mandibles; two small spots behind the ocelli, a large line behind the eyes, a broad line on the side of the pronotum, a mark on the side of the scutellum, metanotum, two large marks on the sides of the median segment, tegulæ, tubercles, a small spot below them, the sides of the petiole, and a short line on the edges of the other segments and on the edges of the third to fifth ventral segments, whitish-yellow. Ocelli not forming a triangle; the outer separated from the eyes by a little more than the length of the third antennal joint and by fully its length from each other; eyes a little converging towards the clypeus. Flagellum rufous beneath, fuscous above; the third and fourth joints sub-equal. Clypeus flat, the apex a little projecting and armed with four stout, blunt teeth. Pronotum rounded at the sides and with an oblique slope in front. Mesopleural furrow indistinct. Trigonal area smooth, impunctate, shining; the central furrow narrow, metapleuræ at base strongly striolated; black above; the rest reddish, the extreme apex of the median segment being also reddish. Petiole rufous, broader than long; the base oblique. Pygidial area reticulated; narrowed almost to a point at the base; the apex transverse. Incision in hypopygium not reaching to the middle, longer than broad, rounded at the base. The basal ventral segment is entirely rufous; the others are broadly rufous in the middle. Wings almost hyaline, the apex infuscated; the second cubital cellule arched, receiving the petiole almost in the middle; the recurrent nervure received a little before the middle of the cellule. Legs rufous in front, more or less whitish-yellow beneath; the hind tibiæ fuscous behind, the hind tarsi for the greater part fuscous. Length 6—7 mm.

The ♂ is smaller and similarly colored, but with the yellow marks more reduced; the clypeus slightly convex; the apex transverse, except that a short, not very distinct tooth projects in the middle. The femora are broadly lined with black laterally and above, especially the anterior four.

A well-marked species.

Cerceris pentadonta. (Pl. X., f. 13, *a, b*.)

Eyes almost parallel. Ocelli in a triangle, the hinder separated from the eyes by nearly the length of the second and third antennal joints united, and from each other by less than the length of the third. Lobe of clypeus obliquely projecting, thick, much broader than long, slightly incised at the apex; the apex with five teeth; the central and outer sharply pointed, the other two broader and truncated at the apex. Third antennal joint not much longer than the fourth and not twice the length of the second. The punctuation on the head is close, moderately strong; the pubescence short and sparse. Thorax opaque, aciculate, and bearing widely separated shallow punctures, those on the median segment more widely separated than those on the mesonotum. Triangular area of median segment opaque, finely and closely rugose, channelled down the middle. The oblique furrow on the mesopleura is wide, deep and obliquely striolated; the metapleuræ finely obliquely striolated; median segment short, its apex obliquely and rather abruptly rounded. Petiole longer than broad, dilated slightly in the middle, clearly separated from the second segment, and covered with moderately large punctures. The other segments are similarly punctured, the punctuation becoming sparser towards the apex. Pygidial area opaque, irregularly and slightly reticulated, much narrowed at the base, the apex rounded. Incision on hypopygium extending beyond the middle, rounded at base, of nearly equal width throughout. The flagellum is rufo-fulvous beneath, the inner orbits of the eyes, the antennal tubercle, the pronotum, metanotum, the third abdominal segment, except at apex, and a band occupying the apical half of the fifth segment, rufo-yellow.* Legs inclining to piceous; the anterior knees, tibiæ and tarsi, yellow in front; the spurs white; the spines blackish. The second recurrent nervure is almost interstitial; second cubital cellule rather shortly pedunculated; the apical nervure longer than the basal. The wings, although, dark smoky throughout, are darker along the radial cellules. Length 8 mm.

---

* I suspect that the rufous tint is caused by cyanide of potassium.

CERCERIS INTERSTINCTA, *Fab.*

What this species may be is rather doubtful. On the whole the description fits best with the ♂ of *humbertiana*, but it differs from it in no mention being made of the two marks on the median segment. With this exception, however, the description, so far as it goes, suits *humbertiana* fairly well.

CERCERIS DISSECTA, *Fab.*

I am quite unable to recognise this species from the description. It may be added that neither Smith nor Schletterer includes this species in their Catalogues.

The description of the Fabrician species are subjoined.

PHILANTHUS INTERSTINCTUS.

P. niger flavo varius abdominis segmento primo rufo. Alis apice nigris.

Habitat in India Dom. Daldorff.

Statura et magnitudo P. ruficornis, antennæ ferrugineæ articulo primo flavo. Caput flavum vertice nigro. Thorax niger, margine antico, puncto calloso ante, lineis duabus scutelli, macula utrinque sub alis, et subscutello flavis. Abdomen nigrum segmento primo rufo, reliquis apice flavis, ano tamen toto nigro, alæ albæ, anticæ apice nigræ. Pedes flavi geniculis nigris.

PHILANTHUS DISSECTUS.

P. niger flavo varius, abdominis basi rufo, macula flava.

Habitat in India Dom. Daldorff.

Praecedente paullo minor, antennæ rufæ, basi flavæ; caput nigrum, labio albo, thorax niger margine antico utrinque lineola, puncto ante alas scutelloque flavis; abdomen nigrum petiolo toto ferrugineo, segmento primo basi ferrugineo, macula flava, secundo quartoque apice imprimis ad latera flavis. Pedes flavi femoribus supra nigris.

B. *Clypeus porret in* ♀.

CERCIRIS ORIENTALIS, *Smith*, (Pl. X., f. 14, *a, b*).

The largest of the oriental species. In the only example I have seen the ocelli are in a black patch; and the sternum is also black. The posterior ocelli are separated from the eyes by fully the length of the second and third joints united; and from each other by the length of the fourth. Trigonal area broad, coarsely aciculate. Petiole broader than long. Pygidial area narrowed a little at base and apex; the apex rounded; the hair-fringe dense pale fulvous; incision in hypopygium reaching near to the middle; there is a tuft of hairs on the apex of the segment at the sides. The second cubital cellule forms almost a semi-circle; and receives the recurrent nervure near the middle.

CERCERIS NEBULOSA, *sp. nov.* (Pl. X., f. 15, *a, b*.)

Eyes slightly diverging beneath. Ocelli hardly forming a triangle; the posterior separated from the eyes by fully the length of the third antennal joint, and by the length of the fourth from each other. Clypeus flat, the middle at the apex turned outwardly, widely semi-circularly incised and separated from the lateral pieces by projecting beyond them. Head closely rugosely punctured; the cheeks and lateral margin of clypeus densely covered with silvery white hair. Mesonotum closely and finely longitudinally rugosely punctured; the pleuræ coarsely rugose; trigonal area finely rugose. Abdomen twice the length of the thorax, and narrower than it; the petiole not much broader than long; the punctuation of the segments not very strong; the pygidial area closely rugose, hardly narrowed at the base; the apex transverse: the incision on hypopygium longer than broad; rounded at the base, dilated at the apex. The third joint of antennæ longer and thinner than the fourth. The cheeks, clypeus, orbits broadly above the base of the antennæ; a spot behind the eyes near the top, a band on the pronotum, scutellum, post scutellum, a narrow band on the base of the second abdominal segment, the extreme apex of the petiole, the apical half of the third segment and the extreme apex of the fourth, reddish-fulvous. Legs black; the anterior tibiæ and tarsi obscure testaceous in front. Length 15 mm.

CERCERIS HUMBERTIANA (Pl. X., f. 16, *a*, *b*.)

In the ♂ the eyes are parallel. The hinder ocelli are separated from them by a little more than the length of the third antennal joint, and by nearly the same distance from each other. The middle of the clypeus projects considerably, the projection almost truncated at the apex, except that there is a waved projection in the middle; the sides of the projection are obliquely curved. The lateral hair fringe is dense and golden. The third antennal joint is a little longer than the fourth and attenuated at the apex. The puncturing on the top of the head is very deep and coarse, on the thorax it is not quite so strong. The trigonal area of median segment is smooth, shining, and impunctuate, and has very narrow and indistinct furrow down the middle. Pygidial area bearing large oval punctures; it is a little narrowed and truncated at the apex and has there a golden pile. The petiole is broader than long and rugosely punctured. The other segments bear large, separated punctures of moderate length.

May be known from the ♂ of *instabilis* by the clypeus of the latter not projecting so squarely in the middle; by the trigonal area being coarsely longitudinally striolated; by the petiole being much narrower, being clearly longer than broad; by the yellow band on the vertex, &c.

Apparently a common species.

CERCERIS EMORTUALIS.

I should say, judging how very variable *humbertiana* is, that this will prove to be a variety of the latter.

CERCERIS HILARIS, *Smith* (Pl. X., f. 17, *a, b.*)

The clypeus is longer than broad, and becomes gradually broader towards the apex; the five basal joints of the flagellum of the antennæ are rufous, and the remaining joints are rufous beneath; the third joint is slightly curved, and about one-third longer than the fourth; the trigonal area is coarsely longitudinally striolate; the hypopygium is transversely rugose, and narrowed gradually towards the apex, and rounded there. Comes nearest to *C. flavopicta*, but is larger, has the clypeus emarginate, the trigonal area striolate, &c.

PSENIDÆ.

1. PSEN RUFIVENTRIS, *sp. nov.*

Niger, mandibulis, pedibus abdomineque rufis; alis clare hyalinis, nervis nigris. ♀. Long. 9 mm.

Antennæ closely covered with a pale microscopic pile; becoming gradually thickened towards the apex; the scape curved, bare; as long as the third joint, which is about a quarter longer than the fourth. Head shining, impunctate above; the front and clypeus closely and finely punctured. Clypeus broadly convex, the apex depressed, gaping and transverse in the middle. Ocelli in pits; a short transverse furrow behind them; the hinder separated from the eyes by about the length of the fourth antennal joint and by a somewhat greater distance from each other. Eyes parallel, coarsely facetted. Mandibles shining, somewhat punctured at the base. The face, cheeks, base of front and of clypeus densely covered with golden hair; the outer orbits of the eyes more sparsely with silvery. Thorax opaque on the mesonotum, the rest shining, impunctate. The apex of median segment irregularly reticulated, laterally striolated. Basal area of median segment semi-circular; depressed at the apex, and bearing some stout keels and with a straight, finely aciculated, shallow furrow in the centre. Pronotum with a slight depression in the middle. The sternum, median segment and pleuræ sparsely covered with long silvery hair. Petiole curved, longer than the thorax. Abdomen shining, impunctate, glabrous. Pygidial area shining, convex, keeled laterally. Legs covered somewhat thickly with white, glistening hair. The second cubital cellule above (the nervures straight) is a little shorter than the space bounded by the first recurrent and second transverse cubital nervures and less than half the length of the third cellule; the recurrent nervures are received about the same distance behind the transverse cubitals. The antennæ are pale fulvous; the mandibles at the base pale testaceous, black at the apex; the legs (except the coxæ) pale ferruginous; the posterior femora and tibiæ a little infuscated; abdomen piceo-ferrugineous, infuscated in the middle.

Psen erraticus, Smith (*Proc. Linn. Soc.* IV., p. 85), from Celebes agrees closely in coloration with the species here described; but no details beyond color are given, and in that respect it differs from *rufiventris* in having the nervures and stigma pale ferruginous, not deep black as in our species.

Hab. Madras (*Rothney*).

PSEN CLAVATUS, sp. nov.

Smaller than *P. rufiventris* (8 mm. only) differing from it in the eyes being rounded in front, diverging at apex, in the antennæ being distinctly clavate; with the third joint not twice the length of the second (in *rufiventris* it is three times) and not much longer than the fourth; in the third joint being nearly twice the length of the penultimate; in there being no furrow behind the ocelli; in the base of the median segment being not depressed, and with three keels down the central part,—one central and two lateral keels; the abdomen apart from the black petiole, is entirely rufous; the second cubital cellule is narrowed almost to a point at the top: the head and thorax are more distinctly aciculate, and there are two shallow furrows on the mesonotum, only the four anterior knees, tibiæ and tarsi are testaceous, this being also the case with the hinder. ♀.

Hab. Poona (*Wroughton*).

—3. PSEN ORIENTALIS, *sp. nov.*

Niger, nitidus, alis fere hyalinis, nervis nigris, ♀. Long. 11 mm.

Antennæ distinctly thickened towards the apex, the flagellum closely covered with a whitish pubescence, the scape thickened, shorter than the third joint, which is twice the length of the fourth. Front and vertex shining, impunctate, sparsely pubescent; the head below the antennæ covered densely with white depressed hair. Ocellar regions raised, the ocelli not in pits. Eyes converging perceptibly towards the bottom. Clypeus almost as in *P. rufiventris*. Thorax shining, impunctate; the basal area of median segment not so clearly defined as in *P. rufiventris*; the keels less distinct; and the apex of the segment is much less distinctly reticulated, and there is a large depression in the centre. Petiole on lower side bearing long white hair. Pygidial area flat above, not convex, the sides keeled, the base impunctate, the rest finely and closely punctured; the sides covered with longish fulvous hair. Legs covered with white hair; the hind spurs and claws ferruginous. The second cubital cellule at the top is shorter than the space bounded by the first recurrent and second transverse cubital nervures; the first recurrent nervure is received in the basal third of the cellule, the second is almost interstitial.

Apart from the difference in coloration, *P. orientalis* may be known from *P. rufiventris* by the third antennal joint being twice the length of the fourth, by the less clearly hyaline wings, by the interstitial second recurrent nervure, &c.

Hab. Madras (*Rothney*).

CRABRONIDÆ.

CRABRO.

I. *Abdomen petiolated.*

1. CRABRO FLAVOPICTA, Smith, *Cat. Hym. Ins.* IV., 391 4.
   Hab. Northern India.[1]

2. CRABRO BUDDHA Cam.
   Hab. Tirhoot, Bombay, Madras.

3. CRABRO ODONTOPHORA, Cam. *postea.*
   Hab. Barrackpore.

II. *Abdomen sessile.*

4. CRABRO ARDENS, Cam., *postea.*
   Hab. Barrackpore (*Rothney*).

5. CRABRO ARGENTATUS, Saint-Fargeau, *Ann. Soc. Ent. Fr.* III., 710, 11 ; *Nat. Hist. Hym. Ins.* III. 194, 1 ;[1] Dahlbom, *Hym. Eur.* I., 385, 255.
   Hab. India.[1]

6. CRABRO FAMILIARIS, Smith, *Proc. Linn. Soc.* II., 106, 1.[1]
   Hab. Borneo[1] (*Wallace*).

7. CRABRO FUSCIPENNIS, St.-Fargeau, *Ann. Soc. Ent. Fr.* III., 710, 11 ; *Nat. Hist. Hym. Ins.* III., 113 ;[1] Dahlbom, *Hym. Eur.* I., 385, 255.
   Hab. India.[1]

8. CRABRO INSIGNIS, Smith, *Cat. Hym. Ins.* IV., 422, 145.[1]
   Hab. India.[1]

9. CRABRO NANUS, Cam. *postea.*
   Hab. Barrackpore.

10. CRABRO RUGOSUS, Smith, *Proc. Linn. Soc.* II., 106, 2.[1]
    Hab. Borneo[1] (*Wallace*).

Hab. Northern India.[1]

CRABRO ODONTOPHORA, *sp. nov.* (Pl. X. f. 20, *a, b*).

Antennæ covered with a microscopic white pile, black, the scape lined with yellow beneath; the third joint nearly twice the length of the fourth. Head with a plumbeous hue, shining, the front, cheeks and clypeus densely covered with silvery pubescence, the outer orbits of the eyes on the lower side with silvery and the vertex with a pale fuscous hair. Ocelli in a triangle, the hinder separated from the eyes by nearly the length of the third and from each other by nearly the length of the fourth antennal joint. A shallow furrow runs down the front from the vertex; the front is only slightly excavated; the clypeus is bluntly keeled, the apex projecting in the middle, almost truncated, but with the sides rounded. Mandibles yellow, black at the apex, covered sparsely with white, glistening hair. Thorax with a plumbeous hue, shining, very minutely punctured on the mesonotum and scutellum; the pronotum narrowed in the middle in front and with a minute furrow on the top; the pleural furrows crenulated; median segment elongate; shining, covered with long white hair; furrowed down the middle, the centre at base finely aciculate, the sides finely rugose; metapleuræ shining, not hollowed, almost impunctate. Abdomen shining, impunctate; the petiole hollowed at the base above; a small mark on either side at its apex; a large mark on the sides of the third and fourth; the apex broadly rounded, almost truncated, but with the sides rounded; pygidium punctured, hollowed in the centre. Legs: hind femora and tibiæ clavate; the latter with a broad furrow on the apical half on the outer side, the former with a broad furrow in the middle on the lower side; basal joint of anterior tarsi twisted, dilated at the base, the dilatation forming a blunt tooth; hind calcaria reaching to the middle of metatarsus; hind coxæ at the base projecting in a stout curved tooth; metatarsus curved at the base beneath and densely pilose, longer than all the other joints united. Hypopygium curved, dilated at the base and armed there with two stout, curved, sharp teeth. The recurrent nervure received considerably before the middle of the radial cellule. Length 7—8 mm.

CRABRO ORIENTALIS, sp. nov.

Scape and second joint of the antennæ clear straw yellow; the other joints black, fuscous beneath, covered with a white microscopic pile; the third joint a very little longer than the fourth. Vertex opaque, finely and closely rugose; covered with a short fuscous-black pubescence; ocelli in a curve, the hinder separated by about the length of the second and third antennal joints united from the eyes, and by a slightly less distance from each other. Front deeply excavated, margined at the top, closely covered with a silvery pile; clypeus slightly convex and with a fine keel down the middle; densely covered with golden pubescence. Mandibles shining, clear yellow, black at the apex. Thorax opaque, alutaceous. The furrow at the base of scutellum crenulated; scutellum longitudinally striolated (at least on the black part), rounded behind; metanotum coarsely longitudinally striolated. Median segment rugosely punctured, irregularly obliquely striolated; the centre furrowed (the furrow at base narrow). Mesopleuræ opaque, alutaceous, the furrows crenulated; metapleuræ finely transversely striated. Petiole fully three-fourths of the total length of the abdomen, the apex nodose, opaque, the base shining, finely punctured. Abdomen opaque, the apical segment punctured. There is a squarish mark on either side of the apex of the petiole; a broad elongated mark on the side of the second segment; a small mark on the side of the third, a broad band on either side of the fourth, narrowed on the inner side and the whole of the fifth segment, yellow. Legs covered with a whitish pubescence; clear yellow; the coxæ and femora to near the apex, and a line on the tibiæ, black; the outer spur of hind tibiæ nearly as long as the metatarsus, broadly dilated. The recurrent nervure is received in the middle of the radial cellule.

May be known from *C. buddha* by there being no shining fovea along the inner orbits of the eyes; by the scutellum being longitudinally striolated; by the petiole and second abdominal segment being marked with yellow, and by the fuscous wings. Length 8 mm.

CRABRO ARDENS, *sp. nov.*

Head shining, finely punctured, less strongly on the vertex than on the front; there is a minute furrow between the ocelli and a shorter one on the outer side; the ocelli hardly form a triangle, the hinder being too close together. Frontal furrow moderately deep and wide. Cheeks and clypeus densely covered with silvery hair; the outer orbits on the lower side not so thickly haired; clypeus keeled in the centre; the sides and apex yellow; mandibles yellow, black at extreme apex. Thorax shining, more strongly punctured than the front, shining; shortly pilose; the sternum densely covered with short pale pubescence; mesopleural furrow wide, crenulated; there is a small round fovea on the posterior lower edge of the mesopleuræ. Pronotum above distinctly raised, transverse, the sides of the raised part oblique; the hinder part of the prothorax separated by a wide and deep depression from the anterior. Scutellum finely punctured, the sides narrowed behind, margined; the basal depression crenulated. Median segment covered with depressed pale hairs; the basal area bounded by a curved furrow, and with some longitudinal striæ; the apical fovea longish, wide; metapleuræ striated.

Abdomen shining, very finely punctured; the sides and apex covered with whitish pubescence; petiole shining, impunctate, depressed at the base; twice longer than wide; pygidial area almost truncate, rounded laterally. Legs at base and tibiæ and tarsi closely covered with a pale pubescence, the anterior broadly bright yellow beneath; the tarsi fuscous-black. Posterior spurs thick, dilated, pale, fully three-fourths of the length of the metatarsus. Antennæ rather densely and longly pilose; the third joint not much longer than the fourth; the scape punctured, yellow in the centre beneath; the base of flagellum fuscous on the lower side. The recurrent nervure is received considerably before the middle of the radial cellule. The tubercles in front and the greater part of the propleuræ are bright yellow.

The ♀ has the oral region and mandibles not yellow; the latter are at the apex brownish; the palpi yellowish; there is a narrow line on the pronotum; the propleuræ are for the greater part black; the front legs are hardly yellow in front; the pygidial area very shining, depressed in the middle, foveate at the apex and indistinctly punctured at the base; the scape is yellow beneath, and the third joint is longer in proportion to the fourth than in the ♂, besides being thinner at the base. Length 5 mm.

CRABRO NANUS, sp. nov.

Head minutely punctured on the front; ocelli in a curve, in pits; a minute furrow in their middle behind; frontal depression wide, deep, shining, a broad keel in the centre; clypeus not carinate, but with the centre broadly projecting at the apex; densely covered with silvery pubescence, as are also the cheeks; outer orbits of the eyes distinctly margined, and bearing a silvery pubescence; the front and vertex bearing a longish blackish pubescence; mandibles yellowish, black at the apex. Scape of antennæ yellow beneath, the flagellum obscure brownish beneath, stout, pilose; the third joint narrow at base, dilated at the apex, not much longer than the fourth. Thorax shining, above very obscurely punctured; the pleuræ, if anything, more distinctly. Pronotum in the middle retreating, the sides broadly rounded, the top not raised, and having a gradually rounded slope to the base; mesopleural suture wide, crenulated, the fovea distinct. There is an indistinct furrow in the centre of the mesonotum, and two foveæ towards the base. Median segment short, the base dilated, the rest with a sharp oblique slope; the basal area irregularly reticulated; the centre is very slightly hollowed, but with the sides of the hollow margined by keels, which unite at the apex; the outer edges are also keeled, the keels converging beneath; metapleural shagreened, and bearing some irregular striae. Abdomen shining, covered almost all over with a pale pubescence; the basal segment, becoming gradually dilated to the apex, which is slightly shorter than the total length; the apical segments rather strongly punctured; the pygidial area rounded at the apex; the ventral segments punctured at the apex. Legs pilose; the long spur of the hind tibiæ reaches to the middle of the metatarsus. The recurrent nervure is received near the middle of the radial cellule. Length 3½ mm.

DASYPROCTUS, *Lep.*

1. DASYPROCTUS CEYLONICUS.

Dasyproctus ceylonicus, Saussure, *Hym. d. Novara Reise*, 85, t. 51.

Hab. Ceylon.

OXYBELUS.

1. OXYBELUS AGILIS, Smith, *Cat. Hym. Ins.* IV., 387, 25.[1]

Hab. India.[1]

2. OXYBELUS ARGENTEOLINEATUS, Cam. *postea.*

Hab. Barrackpore (*Rothney*).

3. OXYBELUS BELLUS, Cam. *postea.*
   Hab. Poona (*Wroughton*).
4. OXYBELUS CANESCENS, Cam. *postea.*
   Hab. Barrackpore (*Rothney*).
5. OXYBELUS FLAVIPES, Cam. *postea.*
   Hab. Barrackpore (*Rothney*).
6. OXYBELUS FULVOPILOSUS, Cam. *postea.*
   Hab. Barrackpore (*Rothney*).
7. OXYBELUS INSULARIS, Kohl, *Természz. Füzetek.* VIII., 109.[1]
   Hab. Ceylon.[1]
8. OXYBELUS NITIDUS, *Cam. postea.*
   Hab. Barrackpore (*Rothney*).
9. OXYBELUS ROBUSTUS, *Cam. postea.*
   Hab. Poona (*Wroughton*).
10. OXYBELUS RUFICORNIS, Smith, *Cat. Hym. Ins.* IV., 388, 27.[1]
    Hab. India.[1]
11. OXYBELUS SABULOSUS, Smith, *Cat. Hym. Ins.* IV., 288 28.[1]
    Hab. India.[1]
12. OXYBELUS SQUAMOSUS, Smith, *Trans. Ent. Soc.*, 1875, 38.
    Hab. Barrackpore (*Rothney*).
13. OXYBELUS TRIDENTATUS, Smith, *Cat. Hym. Ins.* IV., 387, 26.[1]
    Hab. India.[1]

A. *Thorax with laminæ.*

OXYBELUS ROBUSTUS, *sp. nov.* (Pl. X. f. 21—21a.)

Black, a line on the pronotum, two marks on the scutellum, the lateral plates, and a band on each side of the abdominal segments 1—4, white; legs red, the coxæ, trochanters, four posterior tarsi and the four posterior tibiæ behind, black; wings clear hyaline. Eyes wide apart, straight, a little diverging beneath. Antennæ dull rufous, the scape more or less fuscous; the third joint a very little longer than the fourth. Head closely punctured, from a little below the ocelli closely covered with pale silvery hair. Ocelli in a curve, the posterior separated from the eyes by the length of the second, and from each other by the length of the second and third antennal joints united. Apex of clypeus transverse; mandibles obscure rufous. Thorax shining, punctured, the pleuræ more closely than the mesonotum, sparsely covered with a short silvery pubescence; the oblique furrow on the mesopleuræ broad and shallow. Scutellum convex, sparsely covered with large round punctures, the centre keeled, but not strongly, the apex incised in the middle. The lateral plates large, curved laterally and ending in a stout, triangular tooth; the central lamina broader than long, the sides rounded, narrowed at base and apex, the apex with a broad, shallow incision. Abdomen more finely and closely punctured than the thorax; the ventral surface with the punctuation much sparser; the segments at the apex narrowly edged with dirty white. Tibiæ and tarsi bearing a silvery white pubescence; the tibial and tarsal spines white. The recurrent nervure is received almost in the centre of the radial cellule. Length nearly 9 mm.

Most nearly related to *O. squamosus*; but that species may be known from it by the eyes being rounded, not straight, and meeting much closer together, by the blackish antennæ, by the central lamina being rufous and longer than broad (see Pl. X., f. 24).

OXYBELUS FLAVIPES, *sp nov*.

Head closely punctured, the face, cheeks and front densely covered with golden pubescence; the clypeus in the middle, the mandibles and trophi yellow, the tips of the mandibles black, piceous-red before the black. Thorax punctured, closely covered with pale pubescence ; a line on the collar, tubercles, tegulæ, the scutellum, except at the apex, clear stramineous-yellow ; the lateral plates of the metanotum pale yellow. Pronotum rounded, the sides obliquely truncated ; scutellum stoutly keeled down the centre ; the central mucor longer than broad, very slightly and gradually dilated towards the apex, which is acutely and somewhat triangularly, incised ; the lateral plates are large, curved outwardly and ending in a sharply triangular point ; median segment with an oblique slope, finely shagreened ; there is a short furrow below the central mucor ; two keels run from the lateral plates, becoming united near the apex of the segment ; and two other keels run from the sides ; metapleuræ finely longitudinally shagreened. Abdomen shining, closely punctured ; a yellow line on the side of the segments ; pygidial area densely covered with stiff depressed golden-fulvous hairs. Wings clear hyaline, the recurrent nervure quite straight, not oblique, and received before the middle of the radial cellule. Length nearly 5 mm.

OXYBELUS CANESCENS *sp nov*. (Pl. X., f. 22.)

Head closely punctured, densely covered with long grey pubescence. Ocelli separated from each other by twice the distance they are from the eyes, which are very slightly diverging towards the clypeus. Mandibles pale yellow, the apex black. Collar projecting in the middle, slightly concave, and with an oblique slope to the head. Thorax densely covered with a greyish pubescence, strongly punctured, the pleuræ almost rugose ; the oblique furrow crenulated ; there is also a longitudinal furrow, and a narrow oblique one runs down from the apex of the fore wings. Lateral laminæ curved, bi-dentate at the apex, the central curved, convex above, slightly wider at the base, more than twice longer than broad, and incised at the apex. Median segment almost perpendicular, transversely striolated; the central keels not reaching to the apex and forming an elongate triangle, with a smooth ovate fovea at its apex; metapleuræ striated. A line on the pronotum (narrowed in the middle) the tubercles, tegulæ, two large square marks on the scutellum, the metanotum, the lateral laminæ, and the apex of the central mucor, pale yellow. Scutellum keeled down the centre. Abdomen minutely punctured, shining, a large fascia on the first segment, an irregular band, narrowed and interrupted in the middle on the second, and a complete band on the third and fourth segments, pale yellow; pygidial area punctured, covered with stiff fulvous, glistening depressed bristle-like pubescence, and sharply pointed at the apex. Ventral segment with scattered punctures, shining. The recurrent nervure is received considerably before the middle of the radial cellule, and is straight, not oblique. Length nearly 7 mm.

OXYBELUS FULVOPILOSUS, *sp nov.* (Pl X., f. 23.)

Head closely punctured. Covered with a fulvous, the clypeus and cheeks sparsely with silvery pubescence. Eyes slightly converging towards the clypeus; margined; the hinder ocelli separated from each other by twice the length they are from the eyes. Clypeus shining, bluntly and distinctly keeled in the middle. Antennæ with the flagellum fulvous beneath; the third joint a little longer than the fourth and attenuated at the base. Thorax closely punctured, covered with a fulvous silvery pubescence, that on the sternum longish. Collar transverse, the sides obliquely truncated; the mesosternal furrows not very distinct; the lateral laminæ of the metanotum bluntly triangular at the apex; the central convex above, nearly twice longer than broad, and somewhat roundly incised at the apex; median segment with a semi-oblique slope, shagreened, the central keels joining before reaching the apex of the segment and united to the lateral by three keels; the base with some irregular keels; the metapleuræ irregularly striated. Abdomen finely punctured, a large bright yellow macula on the sides of segments one to five, the apical segments punctured, densely covered with stiff rufo-fulvous bristle-like pubescence; the apex rounded. The recurrent nervure is straight and is received a little beyond the basal third of the radial cellule. Length nearly 7 mm.

B. *Thorax without laminæ.*

Oxybelus bellus, *sp. nov.*

Black, closely punctured and covered with a pale fulvous pubescence, a narrow line on the pronotum, the tubercles, a large broad mark below them, the tegulæ in part, a mark on each side of the scutellum, the metanotum, a large mark on the side of the basal two segments, a narrower one on the following two and the legs, clear red; the anterior coxæ black, the four anterior lined with black above, the posterior black, except at the base and apex. Front excavated, densely covered with longish silvery-fulvous hair; this being also the case with the clypeus. Ocelli in a curve, the hinder separated from the eyes by the length of the second antennal joint. Mandibles rufous and yellow, black at the apex. The punctuation on the head is rugose. Thorax strongly punctured; the scutellum with scattered punctures; the metanotum impunctate and without laminæ; median segment reticulated. Abdomen closely punctured; the apical segments covered with longish pale silvery hair; the pygidial area a little longer than broad, rounded at the apex. The anterior tibiæ incline to yellow behind; the legs covered with a short close white pubescence; tibiæ serrate. The eyes are rounded inwardly, converging towards the apex. Wings hyaline, the nervures fuscous; the recurrent nervure is received slightly beyond the middle of the radial cellule. Tibial spurs slender, curved, reaching to the middle of the metatarsus. Abdominal segments lined with silvery pubescence. Length nearly 8 mm.

May be known from *O. argenteolineatus* and *nitidus* by the reddish legs and by the longer and thinner calcaria.

OXYBELUS NITIDUS, sp. nov.

Head smooth, shining, covered with a silvery pubescence, very densely on the clypeus and cheeks, the front excavated in the middle, glabrous. Ocelli in a curve, the hinder separated from each other by only a little greater distance than they are from the eyes. A curved, oblique shallow depression runs from the outer ocelli to the eyes. Clypeus broadly convex, yellow; mandibles yellow, piceous-red at the apex, eyes hairy. Antennæ densely covered with silvery pubescence; the scape yellow, the third and fourth joints equal. Thorax smooth, shining, covered with soft pale pubescence; the pronotum transverse, rounded at the sides. Mesopleural furrow crenulated; medium segment at the base, with short stout keels, bounded by a narrow keel which proceeds to the apex, the sides of the segment being also keeled. Metapleuræ excavated at the base, smooth and shining. There is a small oblong fovea on the metapleuræ behind, a line on the pronotum, tegulæ, tubercles, a large mark below them, a mark on either side of the scutellum, a line on its hinder edge at the sides behind, and a narrower line in the same place on the metanotum, pale yellow. Abdomen shining, smooth; a small mark on the side of the first segment, a large one on the second, none on the third, a large broad one on the fourth, and a smaller one on the fifth, pale pellow; pygidium reddish at the apex, densely covered with soft pale pubescence and punctured. Length 7 mm.

Oxybelus argenteolineatus, sp. nov.

Head strongly punctured, closely covered (the vertex sparsely) with a white silvery pubescence; the mandibles yellow at the base; deeply furrowed. Front ocellus in a shallow depression; an elongated fovea on the outer side of the posterior near the eyes. Antennæ black, shortly pilose, stout; the third joint distinctly longer than the fourth. Thorax close punctured, the pleuræ almost rugosely; the pronotum above, the tubercles and metanotum yellow. Pronotum above transverse, projecting, the sides obliquely curved. Metanotum without laminæ. Median segment short, semi-perpendicular; the base with short stout keels, becoming longer towards the middle, margined, and with two keels which run down to the apex of the segment, converging as they do so. Metapleuræ coarsely shagreened, the base hollowed, shining, and impunctate. Abdomen punctured; the segments lined densely with silvery pubescence, which becomes narrowed in the centre; pygidial area densely covered with stiff rufofulvous depressed pubescence. Legs densely covered with silvery pubescence; the tibiæ more or less lined with black behind; the short stumpy teeth on the hind tibiæ rufous. The recurrent nervure is received in the middle of the radial cellule, and is slightly curved and oblique. Length 6 mm.

Oxybelus Lewisi, sp. nov.

Black, densely covered with a short silvery pubescence; the punctuation close, moderately strong; the mandibles, tubercles, tegulæ in part, two elongated marks on the scutellum, metanotum, a line running from the scutellum and metanotum to the wings, a lateral mark on abdominal segments 1—5 and the knees, tibiæ and tarsi in front, yellow, wings clear hyaline, the nervures pale fulvous. Ocelli in a curve; the hinder separated from the eyes by the length of the third antennal joint; eyes rounded inwardly, converging beneath; the pubescence on the head in front long and dense; the punctuation finely rugose. Scutellum above flat, laterally obliquely margined; the apex of metanotum rounded, without laminæ. Median segment above laterally obliquely, in the centre below, transversely striolate, the striations wide apart and stout; shining, impunctate, metapleuræ finely rugose, obliquely striated; the spine at the base of the median segment stout, acute. Abdomen shining, aciculate; densely pilose; pygidial area acutely pointed, the pubescence fulvous at the apex, dense; the base of the petiole depressed, striolated. The recurrent nervure is received somewhat beyond the basal third of the radial cellule. Length 6 mm.

Hab. Nugata, Ceylon (*George Lewis*).

Explanation of Plates.

PLATE IX.

FIG.
1. *Tachytes Rothneyi*, Cam.
2. *Notogonia luteipennis*, Cam.
3. *Tachytes tarsatus*, Smith.
4. *Ammophila violaceipennis*, Cam.
5. *Notogonia erythropoda*, Cam.
6. *Gastrosericus Wroughtoni*, Cam.
7. *Sphex xanthoptera*, Cam.
8. *Piagetia fasciatipennis*, Cam.
9. *Rhinopsis ruficornis*.
10. *Sphex diabolicus*, Sm. ♂ organs.
11. *Ammophila tyrannica*, Cam.
12. *Sphex aurulenta*, ♂ organs.
13. *Nysson rugosus*, head, 13^a antenna.
14. *Gorytes orientalis*, head, 14^a antenna.
15. Do. *amatorius*, head, 15^a antenna.
16. *Ampulex compressa*, Fab.
17. *Piagetia ruficornis*, Cam. face.
18. *Nysson erythropoda*, head, 18^a antenna.

PLATE X.

FIG.
1. *Stizus reversus*, antenna ♂.
2. *Cerceris viligans*, head, (a) apical ventral, (b) apical dorsal segment.
3. *C. rothneyi*, head, (a) apical ventral, (b) dorsal segment.
4. *C. instabilis*, head and clypeus, (a) apical ventral, (b) dorsal segment.
5. *C. pulchra*, head, (a) apical ventral, (b) dorsal segment.
7. *C. tristis*, head, (a) apical dorsal, (b) ventral segment.
7*. *C. vischnu*, head and clypeus, (a) apical ventral, (b) dorsal segment.
8. *C. novaræ*, head, (a) apical ventral, (b) dorsal segment.
9. *C. Wroughtoni*, head, (a) apical ventral, (b) dorsal segment.
10. *C. dentata*, mesosternum.
11. *C. sulphurea*, head, (a) apical ventral, (b) dorsal segment.
12. *C. tetradonta*, head and clypeus, (a) apical ventral, (b) dorsal segment.
13. *C. pentadonta*, head and clypeus, (a) apical ventral, (b) dorsal segment.
14. *C. orientalis*, head and clypeus, (a) apical ventral, (b) dorsal segment.
15. *C. nebulosa*, head and clypeus, (a) apical ventral, (b) dorsal segment.
16. *C. humbertiana*, head and clypeus, (a) apical ventral, (b) dorsal segment.
17. *C. hilaris*, head and clypeus, (a) apical ventral, (b) dorsal segment.
18. *C. flavopicta*, head and clypeus, (a) apical ventral, (b) dorsal segment.
19. *Philanthus depredator*, head, (a) apical ventral, (b) dorsal segment.
20. *Crabro odontophora*, (a) fore leg, (b) hind leg.
21. *Oxybelus robustus*, squama.
22. Do. *canescens*, squama.
23. Do. *fulvopilosus*, squama.
24. Do. *squamosus*, squama.

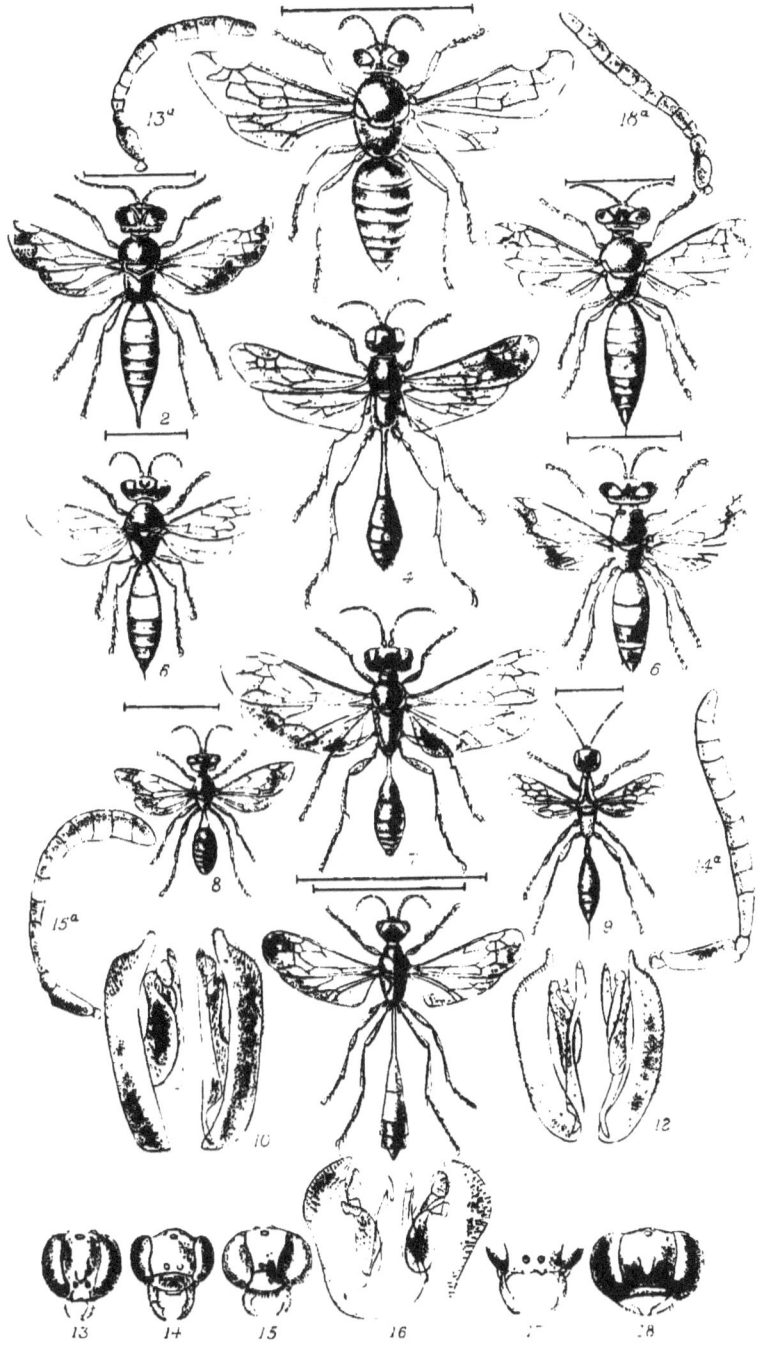

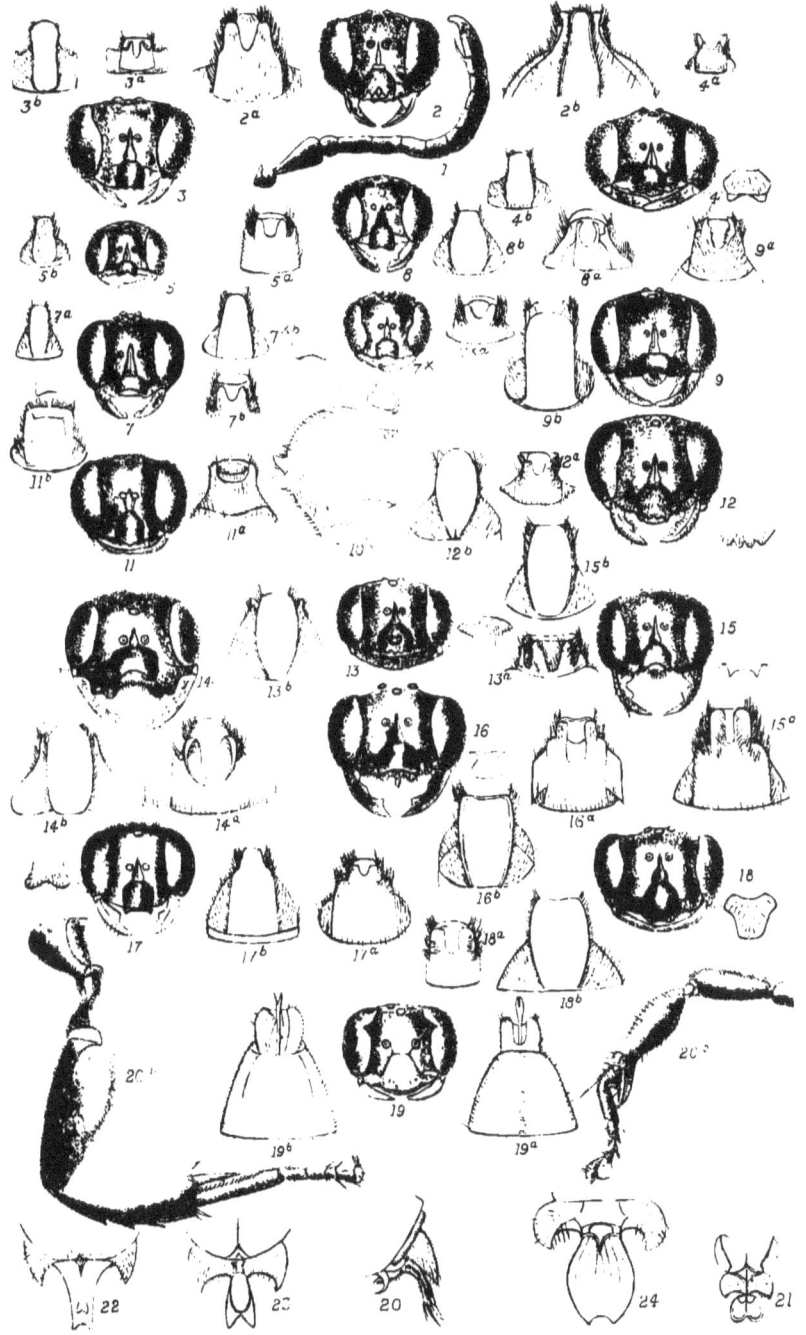

[*From the Fourth Volume of the Fourth Series of* "MEMOIRS AND PROCEEDINGS OF THE MANCHESTER LITERARY AND PHILOSOPHICAL SOCIETY." *Session 1890-91.*]

Hymenoptera Orientalis; or Contributions to a knowledge of the Hymenoptera of the Oriental Zoological Region. III.

BY

P. CAMERON.

MANCHESTER
36, GEORGE STREET.

1891

Hymenoptera Orientalis; or Contributions to a knowledge of the Hymenoptera of the Oriental Zoological Region. By P. Cameron. Communicated by John Boyd.

Received May 1st, 1891.

PART III.

POMPILIDÆ.

I have experienced considerable difficulty in identifying the numerous species of this large family, described by the late Mr. F. Smith, of the British Museum. This is more particularly the case with the black species, and with those related to *Salius flavus*, Fab. These latter I find to be especially puzzling, from the fact that the same type of colouration is found in two of the sections of *Salius* and in *Pompilus*. I have myself, with the aid of numerous examples, come to definite conclusions as to the limits of the species with those of the *flavus*-colouration; but I am in so much difficulty about the nomenclature, that I have decided to leave them over until I have had an opportunity of examining Smith's types. I am the more inclined to do so from finding in Mr. Rothney's collection a *Pompilus*, and a *Salius* named *dorsalis*, Lep., by Mr. Smith.

As regards the genera, I have adopted them as defined by Kohl in his paper "Die Guttungen der Pompiliden" in *Verh. z.-b. Ges. Wien*, 1884.

The species of *Pompilidæ*, as a rule, store their nests with spiders; but very little is known about the habits of the Indian species. Major Bingham describes the nest of *Pompilus bracatus* as a "burrow in the ground at the foot of a large fern," and he observed it provisioning its nest

A I

with a small cockroach. *P. Greeni* was reared by Mr. Green from a cocoon in what had evidently been a large spider's nest between two leaves; and he surmised that the grub had been feeding on the spider's eggs.

CEROPALES.

Ceropales, Latreille, *Prec. caract. gen. Ins.* 1796, p. 123; Kohl, *Verh. z.-b. Ges. Wien*, 1884, p. 51.

1. CEROPALES FUSCIPENNIS.

Ceropales fucipennis, Smith, *Cat. Hym. Ins.*, iii., p. 179.[1]
Hab. India.[1]

This agrees with *Orientalis* closely in the colouration of the head, thorax, and legs, but differs in having the abdomen ferruginous, black at the base.

2. CEROPALES ORIENTALIS, *sp. nov.* (Pl. III. f. 4).

Black, pruinose, the abdomen with a bluish tinge; the clypeus, except a triangular black mark in the centre, the mandibles except at the apex, the inner orbits broadly to the ocelli, the outer orbits from near the top of the eyes to the mandibles, a broad line on the pronotum, narrowly incised in the middle, and broadly at the sides in front, two elongate marks on the scutellum, a small mark on the propleuræ, a large one immediately over the middle coxæ; the fore coxæ broadly beneath, the middle coxæ with a small and a large mark beneath and two broad bands on the base of the second abdominal segment, clear whitish-yellow; the trochanters, femora, tibiæ, and tarsi red; the spurs and a line on the four hinder tibiæ yellow; the middle tibiæ entirely, the hinder black and yellow behind; the tarsal joints black at the apices, the anterior with the joints whitish at the base. Eyes slightly curved above, reaching to the base of the mandibles, diverging slightly beneath. Ocelli in a curve, the vertex depressed in front of them, and a minute furrow runs down from them to the antennæ; the hinder separated by a somewhat greater distance from each other than they are from the eyes. Clypeus at the apex margined, forming a rounded curve. Antennæ stout, brownish beneath, the joints curved beneath; the third joint a little shorter than the fourth. Pronotum obliquely rounded at the sides. Metanotum densely covered with long whitish hairs; median segment with a bluish tinge; and a gradually-rounded slope; alutaceous, sparsely covered with long white hairs. There is a longitudinal furrow in the centre of the mesopleuræ. Abdomen sessile, granular, the basal segment covered thickly with stout white depressed hair, the rest of the abdomen pruinose. Legs thickly pruinose; the tibiæ sparcely spined; the long spur of the hind tibiæ fully two-thirds of the length of the metatarsus. Wings iridescent, hyaline at the base; from the basal nervure suffused with dark fuscous, darkest at the apex; the second cubital cellule at the top and bottom a little longer than the third; the nervures blackish.

Length 10 mm.

Hab. Barrackpore (*Rothney*).

4. CEROPALES CLARIPENNIS, *sp. nov.*

Black, shining, the mandibles, clypeus, face, orbits, except narrowly interrupted at the top; a line on the pronotum behind, the angles in front, a broad line at the apex of the mesonotum, a line on the scutellum, the metanotum, a narrow line down the middle of median segment; the propleuræ beneath, a broad oblique band on the mesopleuræ above and two smaller ones on the lower half, and two large marks on the metapleuræ, clear yellow. Abdomen ferruginous, the extreme base black. Legs reddish; the fore coxæ yellow, with a black mark behind; the four hinder coxæ black, yellow and red beneath, the tarsi black; the spurs pale, the hinder about two-thirds of the length of the metatarsus. Wings clear hyaline, the second cellule half the length of the third above and beneath; the first recurrent nervure received slightly past, the second slightly in front of the middle of the cellules. Antennæ moderately thick; brownish beneath at the base; the third and fourth joints subequal. Eyes with a distinct curve at the top, distinctly converging at the apex; they being there separated by a little more than half the length they are at the top. Clypeus with the sides oblique, the apex transverse. Ocelli in a triangle, separated from the eyes by twice the length they are from each other. Pronotum semi-transverse behind, quadrate; the sides at the base projecting into triangular teeth. Mesonotum flattish, with two longitudinal furrows; the furrow on the mesoplurae narrow; metanotum gibbous; median segment with a gradual slope. Except on the median segment the body is almost glabrous.

Length, 8—9 millim.

Hab. Poona (*Wroughton*).

5. CEROPALES FLAVOPICTA.

Ceropales flavopicta, Smith, *Cat. Hym. Ins.* III., p. 178, 5[1].
Hab. India.

6. CEROPALES ORNATA.

Ceropales ornata, Smith, *Cat. Hym. Ins.* III., p. 179[1].
Hab. India.

7. CEROPALES ANNULITARSIS, *sp. nov.*

Yellow, a stripe across the vertex behind the eyes, a broad one leading down from it on the front, with a small yellow mark on its centre, a broad band in front of the pronotum, from which an oblique one runs up the pleuræ, the mesonotum, except along the sides, and a large squarish mark in the centre, this latter having a large black triangular mark in the centre, the mesosternum, a large oblique mark on the base of the mesopleuræ, the base of the scutellum and the metanotum, black ; the greater part of the meso- and meta- pleuræ and the median segment, reddish. Abdomen yellow; the first segment black at base and apex, the centre reddish ; the second segment black and red at the apex ; the third broadly black : the fourth black at the apex, the black in the middle being continued to the base of the segment ; the fifth black at the base, the black being continued in the middle to the apex ; the ventral segments broadly black. Legs ferruginous ; the coxæ yellow and red ; the trochanters blackish ; the apex of the hinder tibiæ and of the four hinder tarsal joints black. Wings yellowish hyaline ; the apex of both wings infuscated ; the second cellule at the top and bottom longer than the third ; the second and third transverse cubital nervures elbowed at the middle, thus narrowiug the second cubital cellule at the top ; the first recurrent nervure is received in the apical third, the second a little before the middle. Antennæ ferruginous, longish ; the joints curved ; the third and fourth joints sub-equal. Apex of clypeus bluntly rounded ; the sides rounded. Head, pronotum, and median segments bearing long white hair. ♂.

Long. 14 mm.

Hab. Poona (*Wroughton*).

How far the ferruginous colour is natural or discoloured by cyanide of potassium, I can't well make out. Certainly some parts of the body are so discoloured.

MACROMERIS, *Lep.*

Lepeletier de Saint-Fargeau, *Guér., Mag. Zool.,* XIV, pl. 29, 1831 ; Kohl, *l. c.* 41.

1. MACROMERIS VIOLACEA, Lep.
Lep., *Nat. Hist. d. Ins. Hym.* III., 463.

A common Indian species.

Hab. Barrackpore, Poona, Madras, Myssoure, China, Malacca, Borneo, Java, Gilolo, New Guinea, Celebes, Key, Aru, Floris.

2. M. SPLENDIDA, Lep.
Lep. *l. c.* 463[1].
Hab. Java[1].

3. M. ARGENTIFEROUS, Smith.
Smith, *Jour. Linn. Soc.* II., 97[1].
Hab. Borneo, Malacca, Singapore, Java[1].

In *Jour. Linn. Soc.*, 1867, p. 556, only Borneo is given as a habitat.

PSEUDAGENIA, *Kohl.*

Verh. z.-b. Ges. Wien, 1884, 38 = *Agenia.* Dbm. non Schiödte, which = *Pogonius*, Dbm.

The basal nervure is said by Kohl to be interstitial; but this is not the case with many of our species.

1. PSEUDAGENIA ÆGINA, Smith, *Proc. Linn. Soc.* II, 94, 9.
Hab. Borneo.

2. P. ALARIS, Saussure, *Hym. d. Novara Reise*, 52[1].
Hab. Ceylon.[1]

3. P. ARIEL, Cam., *postea.*
Hab. Barrackpore *(Rothney).*

4. P. ATALANTA, Smith, *Proc. Linn. Soc.* II., 94, 8[1].
Hab. Borneo, Singapore, Malacca, Bachian, Celebes[1].

5. P. BIPENNIS, Saussure, *Hym. d. Novara Reise*, 52[1].
Hab. Ceylon[1].

6. P. BLANDA, Guérin, *Voy. d. Coq.* II., 260; Smith, *Proc. Linn. Soc.* II., 94, 7.
Hab. India, Malacca, Borneo, Celebes, Ceram, Key, Flores[1].

7. P. CÆRULEA, Smith, *Cat. Hym.*, III., 147, 141.[1]
Hab. India.[1]

8. P. CELAENO, Smith, *Proc. Linn. Soc.*, II., 96, 15.[1]
   Hab. Singapore.[1]

9. P. CONCOLOR, Saussure, *Hym. d. Novara Reise*, 5, 4.[1]
   Hab. Ceylon.[1]

10. P. DAPHNE, Smith, *Proc. Linn. Soc.*, II., 95, 10.[1]
    Hab. Borneo.[1]

11. P. FLAVOPICTA, Smith, *lc.*, 96, 13.[1]
    Hab. Singapore.[1]

12. P. FESTINATA, Smith, *Trans. Ent. Soc.*, 1875, 37.
    Hab. Barackpore (*Rothney*).

13. P. FRAUNFELDIANA, Saussure, *Hym. d. Novara Reise*, 53,[1] f. 35.
    Hab. Java, Batavia.[1]

14. P. HIPPOLYTE, Smith, *Proc. Linn. Soc.*, II., 96, 14.[1]
    Hab. Singapore.[1]

15. P. INSULARIS, Saussure, *Hym. d. Novara Reise*, 55.
    Hab. Ceylon.

16. P. LAVERNA, Smith, *Proc. Linn. Soc.*, II., 95, 1.[1]
    Hab. Borneo.

17. P. MACULATA, Taschenberg, *Zeits. f. Gess. Wissen.*, 45, 1.
    Hab. Java.

18. P. MELAMPUS, Smith, *Proc. Linn Soc.* II., 95, 12[1].
    Hab. Borneo.[1]

19. P. MICROMEGAS, Saussure, *Hym. d. Novara Reise*, 51, *fs.* 35 *a-b*.
    Hab. Ceylon.[1]

20. P. MUTABILIS, Sm.,
    Agenia mutabilis, Smith, *Trans. Linn. Soc.* VII. 186.[1]
    Hab. Mainpuri, North-West Provinces.

21. P. TINCTA, Smith, *Cat. Hym.* III. 145, 152.[1]
    Hab. India.[1]

22. P. VARIPES, Dahlbom, *Hym. Eur.* I. 455, 7.[1]
   Hab. India.

23. P. NANA, Saussure, *Hym. d. Novara Reise*, 55.[1]
   Hab. Ceylon.[1]

24. P. OBSOLITA, Saussure, *l. c.* 56, f. 37.[1]
   Hab. Ceylon.[1]

25. P. PLEBEJA, Sauss.
   Saussure, *Hym. d. Novara Reise*, 57.[1]
   Hab. Ceylon.

26. P. VEDA, Cam. postia.
   Hab. Poona (*Wroughton*).

PSEUDAGENIA CAERULEUS, *Smith.*

A specimen from Barrackpore is probably this species; but the description is rather incomplete. The clypeus at the apex is broadly rounded, the sides obliquely truncated; the ocelli in a triangle and separated from the eyes by a somewhat greater distance than they are from each other; occiput slightly concave; sides of pronotum rounded; shorter than the head; second and third cubital cellules subequal; the first recurrent nervure received shortly before the middle of the cellule; the second about half the length of the transverse cubital nervure from the base. I cannot see the "fuscous cloud transversing the externo-medial nervure," nor "a faint cloud" in the second submarginal cellule. The apex of the hind femora is black, a fact not mentioned by Smith for his *cærulea*. It is possible that my specimen may be *cyaneus*, Lep., but that has the third cubital cellule "plus grande que la deuxième." The median segment at the apex rounded, transversely striated, and having a gradually-rounded slope to the apex; the apex also having a tuft of white hair on either side; the upper part of metapleuræ obliquely striated; the long spur of the hind tibiæ does not reach the middle of the metatarsus; the base of the latter with a thick hair brush; the other joints with short spines beneath; the fore tarsi pilose beneath.

PSEUDAGENIA FESTINATA, *Smith*. (Pl. III. f. 3).

This species, I consider, identical with *P. alaris*, Sauss. Smith's type is smaller, and the wings have not the yellowish tint quite so marked.

PSEUDAGENIA CELÆNO, *Smith*.

A ♂ from Barrackpore, is, perhaps, this species—at least it agrees fairly well with the description so far as that goes. The eyes distinctly converge towards the apex; the clypeus is transverse at the apex; the sides being oblique; the ocelli form a triangle, and are separated from each other by a perceptibly less distance than they are from the eyes; the second and third cubital cellules at the top and bottom are subequal; the first recurrent nervure is received a little before the middle; the second in the basal third of the cellule; the nervures are pale testaceous. From *alaris* it is easily known by the truncated apex of the clypeus. The long spur of hind tibiæ does not reach the apex; the metatarsal brush slight.

PSEUDAGENIA ARIEL, *sp. nov.*

Black, shining, pubescent, eyes distinctly converging beneath, the space separating them at the top being distinctly greater than at the bottom. Ocelli in a triangle; the hinder separated from each other by a less distance than they are from the eyes. Clypeus convex, the basal half laterally oblique; the apical curved, terminating in a blunt point. Occiput bluntly rounded. An indistinct furrow runs from the ocelli to the antennæ. The head is convex in front, shining, finely punctured, sparsely covered with long silvery hairs; the cheeks and clypeus bear a silvery pubescence; antennæ longish, stout, pruinose, tapering towards the apex. Mandibles at the base finely rugose. Pronotum broad, broadly rounded in front, behind concave. Median segment short, with a rather abrupt slope, transversely striated. The thorax above in front is shining, minutely punctured, laterally opaque, alutaceous, the median segment transversely striated. Abdomen shining, pruinose; the apical segment above shining, impunctate, sparsely covered laterally with long pale hairs. Wings sub-hyaline; the second cubital cellule above slightly longer than the third; below a little shorter; the first recurrent nervure is received a little before the middle; the second in the basal third. Legs pruinose; the long spur of the hind tibiae does not reach the middle; tarsi with a few fulvous spines.

PSEUDAGENIA VEDA, sp. nov.

Black, wings clear hyaline; a small fuscous cloud below and touching the stigma. Eyes a little converging, the hinder ocelli separated from each other by a very slightly less distance than they are from the eyes. Clypeus short, convex, the apex broadly rounded. Occiput slightly concave in the middle. The front strongly aciculate; the vertex shining, almost impunctate. The head, except on the vertex densely covered with a silvery pubescence; the vertex with a few fuscous hairs; the lower and outer orbits with some long silvery ones. Pronotum shorter than the head, roundly narrowed towards the head, shallowly concave behind. Pro- and meso- thorax alutaceous; the median segment with a rounded slope, irregularly transversely striated; deeply furrowed down the centre, the sides covered with long whitish hairs. Abdomen shining; pruinose, having an olive tint, the petiole with a distinct neck. Radial cellule wide, angled where the cubital nervures are received; the second cubital cellule at the top distinctly shorter than the third, especially on the lower side; the first and second transverse cubital nervures with a slight oblique curve; the first recurrent nervure is received a little beyond the middle; the second at a less distance from the base. Legs densely covered with a silvery pile; the long spur of the hind tibiæ not much more than a third of the length of the metatarsus; the front spurs pale, the front tarsi fuscous; the tibiæ with short spines; the metatarsal brush slight.

This species differs from the others in having the basal nervure interstitial; but in other respects it agrees with the generic character.

Length 7mm.

PSEUDAGENIA TINCTA, *Smith*.

Black, densely pruinose, the hinder femora red, wings hyaline, the nervures black. Eyes curved, converging beneath and at top. Ocelli separated from the eyes by a somewhat greater distance than they are from each other. Clypeus broadly convex, broadly rounded at the apex. Occiput slightly convex, the sides rounded. A shallow furrow runs down from the ocelli. Head opaque, alutaceous; the clypeus and cheeks densely covered with a silvery pubescence; the front and vertex sparsely silvery pubescent, shewing fuscous hairs. Antennæ stout, involute. Thorax alutaceous, covered with a silvery pubescence; the pleurae and median segment with long soft white hairs; the pronotum short, bulging out roundly laterally, behind slightly curved, angled in the middle. Median segment obscurely transversely striolate, indistinctly channelled down the centre. Petiole with a distinct neck, becoming gradually widened towards the apex; the abdomen shining, laterally, and at the apex densely pruinose; the apical segment impunctate, the apex with fulvous hairs. Radial cellule elbowed slightly where the first and third transverse cubital nervures are received; the second cubital cellule a little shorter than the third at top and bottom;

the first recurrent nervure is received a little in front of the middle of the cellule; the second in the basal third. Legs elongated, pruinose; the long spur of the hind tibiæ does not reach the middle of the cellule. The abdominal segments are testaceous at the apex.

It is probable that this is *P. tinctus*, Smith, with which it agrees in colouration (the only point noted in the description), except that I fail to notice any trace of "green tinge" about the head and thorax. The four anterior femora may be entirely black, or more or less reddish.

It is probably *varipes*, Dbm.

SALIUS, *Fab.*, sec. *Kohl.*

Salius, Fab., *Syst. Piez.*, 124; Kohl, *Verh. z-b, Ges. Wien*, 1884, 43.

Procnemis, Schiödte, *Mon. Pomp. Kröyer Tidsskr.*, I, 1837.
Hemipepsis, Dbm., *Hym. Eur.*, I., 462.
Homonotus, Dbm., *l. c.* 441, *pt.*
Entypus, Dbm., *l.c.*, 442.
Pallosoma, Smith, *Cat. Hym.*, III., 181.

The following species are in all probability referrable to *Hemipepsis*.

1. SALIUS ÆRUGINOSA, Sm.
Mygnimia æruginosa, Smith, *Cat. Hym.* III., 184, 8[1].
Hab. Sumatra[1].

2. S. ALBIPLAGIATA.
Mygnimia albiplagiata, Smith, *l.c.* 183, 6[1].
Hab. Java[1].

S. AUDAX, Smith.
Mygnimia audax, *l.c.* 182, 4[1]; Bingham, *Journ. Bourb. Nat. Hist. Soc.* V. 239[2].
Hab. Silhet[1]; Kumaon[2].

3. S. ANTHRACINA, Sm.
Mygnimia anthracina, Smith, *l.c.*, 183, 5.[1]
Hab. Malacca, Borneo, Singapore, Sumatra.[1]

4. S. AUREOSERICEA, Guér.
   Pompilus aureosericus, Guér., Voy. Coq., II., 256.[1]
   Mygnimia aureosericea, Smith, Cat. Hym., III., 182, 3.[1]
   Hab. Java.[1]
5. S. AVICULUS, Sauss.
   Mygnimia avicula, Saussure, Novara Reise, 64, fig. 28.[1]
   Hab. Java.[1]
6. SALIUS BELICOSUS, Sm.
   Mygnimia belicosa, Smith. Ann. Mag. Nat. Hist., 1873, 256.[1]
   Hab. Bengal.[1]
7. SALIUS CEYLONICUS, Sauss.
   Mygnimia ceylonica, Saussure, Novara Reise, 64.[1]
   Hab. Ceylon.[1]
8. S. CYANEUS, Lep.
   Pallosoma cyanea, Lep., Nat. Hist. Hym. Ins. III., 493.[1]
   Hab. Java.[1]
9. S. DUCALIS, Sm.
   Mygnimia ducalis, Smith, Proc. Linn. Soc., II., 983.[1]
   Hab. Malacca, Sumatra.[1]
10. S. EXCELSUS, Cam.
    Mygnimia atropos, Smith, Trans. Ent. Soc, 1875, 36, non., Smith, 1855.
    Hab. Barrackpore[1] (Rothney).
11. S. FLAVUS, Fab.
    Pompilus flavus, Fab., Syst. Piez., 197, 2; Lep. Nat. Hist. Hym. Ins., III., 430, 2.[1]
    Sphex flava, Drury, Ill. Exot. Ins., III. t., 42, f. 4, ♀.
    Hemipepsis flavus, Dbm., Hym. Ent., I., 123.
    Hab. Borneo, Singapore, Gilolo, Sumatra (teste Smith).
12. S. FLAVICORNIS, Fab.
    Pepsis flavicornis, Fab. Syst. Piez. 216, 44[1].
    Hab. Malabar.[1]

13. S. FUNESTUS, Cam.
    Mygnimia fenestrata, Smith, *Cat. Hym. Ins.*, III. 184, 10[1] (*non* Smith, *l. c.* p. 147).
    Hab. Silhet[1]

14. S. GROSSUS, Fab.
    Pepsis grossa, Fab., *Syst. Piez.* 214, 32[1]; Lep. *Nat. Hist. Hym. Ins.*, III. 487; Dbm. *Hym. Eur.* I. 464.
    Hab. India.[1]

15. S. HERCULES, Cam. *postea*.
    Hab. Naga Hills.

16. S. FULVIPENNIS, Fab. (Pl. III. f. 28).
    Sphex fulvipennis, Fab., *Ent. Syst.* II., 218, 84.
    Pompilus fulvipennis, Fab., *Syst Piez*, 189, 57.
    Hemipepsis fulvipennis, Dbm., *Hym. Eur.* I., 462, 2.
    Pompilus fulvipennis, Saussure, Novara Reise, 58[1].
    Hab. Madras (*Rothney*), Ceylon[1].

17. S. INDICUS, Cam. postea.
    Hab. Tavoz.

18. S. INTERMEDIUS, Smith.
    Mygnimia intermedia, Smith, *Ann. Mag. Nat. Hist.*, 1873, 257[1].
    Hab. North India[1], Ceylon.[1]

19. S. IRIDIPENNIS, Smith.
    Mygnimia iridipennis, Smith, *Proc. Linn. Soc.* II. 98, 5[1].
    Hab. Malacca[1], Borneo[1], Ceram[1], Timor[1].

20. S. LUSCUS, Fab.
    Pepsis lusca, Fab., *Syst. Piez.* 215, 38[1].
    Priocnemis luscus, Dbm., *Hym. Eur.*, I., 45. 7[2].
    Hab. Tranquebar[1], Port Natal[2] (?).

21. S. LÆTA, Sm.
    Mygnimia læta, Smith, *Ann. Mag. Nat. Hist.*, 1873, 256.[1]
    Hab. Burma.[1]

22. S. MEGÆRA, Cam.
   Mygnimia perplexa, Sm.. *Cat. Hym.*, III., 185, 11[1] (*non l.c.*, p. 147).
   Hab. Madras.[1]

23. S. MOMENTOSUS, Sm.
   Mygnimia momentosa, Sm., *l.c.*, 258.[1]
   Hab. Borneo.[1]

24. S. NIGRITUS, Lep.
   Pallasoma nigrita, Lep., *Nat. Hist. Hym. Ins.*, III., 493.[1]
   Hab. Java.[1]

25. S. PRINCEPS, Sm.
   Mygnimia princeps, Smith, *Proc. Linn. Soc.* II. 98, 4.[1]
   Hab. Borneo.

26. S. PURPUREIPENNIS, Smith.
   Mygnimia purpureipennis, Smith, *Ann. Mag. Nat. Hist.*, 1873, 258.
   Hab. Java.[1]

27. S. RUBIDA, Bing.
   Mygnimia rubida, Bingham, *Jour. Bomb. Nat. Hist. Soc.* V. 238.[1]
   Hab. Ceylon.

28. S. SÆVISSIMA, Sm.
   Mygnimia sævissima, Smith, *Ann. Mag. Nat. Hist.*, 1873, 256.[1]
   Hab. Bombay Presidency.[1]

29. S. SEVERUS, Drury.
   Sphex severus, Drury, *Ill. Exot. Ins.*, III., t. 42, f. 4.[1]
   Mygnimia severa, Smith, *Cat. Hym. Ins.*, III., 182, 1.
   Hab. India.[1]

30. S. VEDA, Cam., *postea*.
   Hab. Poona (*Wroughton*).

31. S. VITRIPENNIS, Sm.

Mygnimia vitripennis, Smith, *Ann. Mag. Nat. Hist.*, 1873, 257.[1]

Hab. Sumatra.[1]

To this section is probably referrable :—

PEPSIS DISELENE, Smith, *Cat. Hym.*, III., 200, 51.

Hab. India, Singapore.

SALIUS (MYGNIMIA) EXCELSUS, *Cam.*

Has the typical *Hemipepsis* wing. Head slightly convex in front and behind. Eyes not arcuate at top, parallel or nearly so; ocelli in a curve; the posterior separated from each other by a distinctly greater distance than they are from the eyes; the ocellar region raised, a depression on either side of it. Clypeus convex, the apex depressed, waved inwardly in the centre. Pronotum a little shorter than the head, the sides rounded, narrowed towards the base; the apex roundly concave. Pronotum short, sharply oblique from base to apex, there being no break in the surface from the base to the apex; the lateral tubercles distinct. Abdomen subsessile; the apex with a thick tuft of hair. The third cubital cellule at the top is a little longer, at bottom considerably shorter than the second; the second recurrent nervure received a little before the middle—a little less than the length of the second transverse cubital nervure; the anal nervure in hind wings interstitial. The long spur of the hind tibiæ reaches beyond the middle of the metatarsus.

There is another larger species (22 mm.) which resembles *excelsus* in colouration, except that the body has a bluish tinge, and the wings have a deep purple iridescence. This is probably *vitripennis*, Sm., from Sumatra. From *excelsus* it is easily known by the median segment having an oblique slope to the apex, when it curves down obliquely; the anal nervure in hind wing is received beyond the cubital— the ocelli are larger.

SALIUS (HEMIPEPSIS) ANTHRACINA, Sm.

This species has the head in front concave, but with the antennal tubercles large, projecting; the eyes parallel from the top; the ocelli large, separated from the eyes by one and a half times the distance the posterior are from each other; the pronotum rounded in front; concave behind: base of abdomen subsessile; apical segment covered with stiff blackish pubescence, roughened. Tibiæ and tarsi thickly spurred; the long spur of the hind tibiæ not one-fourth the length of the metatarsus; thick, obliquely narrowed towards the apex; claws with two stout spines. In the hind wings the anal-nervure is received a little beyond the cubital. Clypeus transverse at apex, projecting.

SALIUS HERCULES, sp. nov.

Black; the face, orbits, tibiæ and tarsi dull brown, the flagellum dull ferruginous, blackish above at the base; wings at the base (to a little beyond the basal nervure), deep blackish-violaceous, the rest brownish-yellow, except at the apex which is infuscated. Head almost transverse in front, behind slightly convex, piceous. Eyes parallel, very little arcuate above; ocelli large, in a curve, the hinder separated from the eyes by a perceptibly less distance than they are from each other. Clypeus moderately convex, below the eyes projecting nearly as much as half the length of the mandibles; broader than long; transverse at the apex; labrum rounded at the apex, nearly half the length of the clypeus. The ocellar region slightly raised, a furrow at the sides of the region, the inner orbits narrowly edged with obscure testaceous; the head thickly covered with a black to fuscous pubescence. Thorax opaque, thickly covered with a blackish pubescence; the prosternum and anterior coxæ with long blackish hairs. Pronotum as long as the head, rounded and narrowed in front, behind convex. Scutellum gibbous, becoming gradually raised and narrowed to the apex which is rounded; metanotum forming a longer tubercle, oblique, haired at the sides, the top rounded, glabrous, and brown. Median segment with a slight straight slope to the apex, which is oblique; transversely striolate; the basal tubercles not very distinct. Metapleuræ shining, impunctate, almost glabrous. Abdomen subsessile, smooth, shining, obscurely pruinose; the apical segments with long hairs. Legs long, stout, the hinder longer than the body, their tibiæ with a row of sharp moderately long spines; the tarsi also spined; the metatarsus with the brush distinct; the long spur of the hind tibiae reaching somewhat beyond the middle of the metatarsus; the tarsal joints blackish at the apex; the claws bidentate. Antennæ stout, bare; the third joint a little longer than the fourth. The second recurrent nervure is received the length of the second transverse cubital cellule from the base of the cellule; the first interstitial; the anal nervure in hind wings interstitial. ♂.

Length 33 mm.

Nearly related to *S. anthracina*, Sm., which differs from it in the ocelli being smaller, and separated from the eyes by a greater distance than they are from each other; in the clypeus being rounded, and in the form of the scutellum, metanotum and median segment.

Hab. Naga Hills.

SALIUS (MYGNIMIA) INDICUS, *sp. nov.*

Black: the antennæ, abdomen and legs ferruginous; the basal half of petiole and coxæ and trochanters, black; the mandibles ferruginous, piceous at the apex; wings deep violaceous. Head transverse before and behind, the eyes projecting beyond the face in front; eyes parallel; at top very slightly arcuate; the ocelli large, in a triangle, the posterior separated from the eyes by a greater distance than they are from each other; a distinct furrow runs down from them. Clypeus with the sides oblique, the apex shining, smooth, slightly arcuate ; apex of labrum with a slight inward curve. Antennæ short, thick, involute ; the third joint nearly twice the length of the fourth. Prothorax shorter than the head, narrowed and rounded towards the base, in front broadly convex. Median segment with a gradually rounded slope to the apex which is almost transverse, transversely striolated ; the sides at the base broadly depressed laterally and with a broad raised margin ; the lateral tubercles elongate. Abdomen subpetiolate, a little longer than the head and thorax united ; acute at the apex, the apical segment with scattered punctures and (especially at the apical half) bearing long black hairs ; beneath it is punctured, except at the base. Legs of moderate length, stout, the tibiæ and tarsi with golden pubescence, the hind tibiæ sparsely spinose ; the long spur of the hind tibiæ does not reach the middle of the metatarsus, metatarsal brush incomplete; the other joints spined. The second recurrent nervure is curved, and is received about the length of the second transverse cubital nervure from the base of the cellule ; at the top the second cubital cellule is much shorter than the third ; at the bottom they are subequal. In the hind wings the anal nervure is received beyond the apex of the cubital.

Length 23 mm.

Hab. Tavoz, *Mus. Cal.*

A well marked species.

SALIUS (MYGNIMIA) VEDA, *sp. nov.*

Black ; the abdomen and legs rufous ; the scape beneath and orbits and the face obscure yellowish ; the flagellum brownish beneath ; wings dark smoky-fuscous. Eyes a little converging beneath ; ocelli separated from each other by a greater distance than they are from the eyes, situated on a raised space ; a narrow furrow surrounding them, and with a depression in front of the anterior. Clypeus broadly convex; the apex rounded; the labrum projecting beyond it; a deepish depression on the sides of the clypeus at the base; occiput transverse in the middle, the sides rounded. Pronotum shorter than the head, a little narrowed anteriorly. Median segment with a slight slope to the apex, when it becomes oblique; apex bluntly rounded. Head and thorax alutaceous, bearing a pale thick whitish pile, the median segment having also some fuscous hairs. Antennæ stout, shorter than the body, the joints dilated beneath; the third and fourth subequal. Abdomen shining, slightly pruinose; the basal 2-joints black. Wings large; the basal abscissa of radius short, a little oblique; the second straight, the third sharply oblique; the first transverse cubital nervure very oblique, the second nearly straight, the third curved, roundly elbowed at the top: the second cubital cellule at the top distinctly shorter than the third, at the bottom fully longer than it: the first recurrent nervure almost interstitial; the second received a little before the middle; the first discoidal nervure bullated; the basal nervure elbowed at the middle. Legs elongate, stout; the tibiæ with a few spines; the coxæ and trochanters black, the greater part of the hind tarsi fuscous; the long spur of the hind tibiæ does not reach the middle of the metatarsus; claws with a tooth in the centre, tibiæ sparsely spined.

Length slightly over 9 mm.

Hab. Poona (Wroughton).

The following are all referrible, no doubt, to *Priconemis*, Kohl's section 2.

30. S. CANIFRONS, Sm.

Pompilus canifrons, Smith, *Cat. Hym.*, III., 146, 138.[1]

Hab. Sumatra[1], Poona (*Wroughton*).

31. S. CONCOLOR, Saus.

Priconemis concolor, Saussure, *Novara Reise*, 54.[1]

Hab. Ceylon.[1]

33. S. CONVEXUS, Bingham.
   Priconemis convexus, Bingham, *Jour. Bomb. Nat. Hist. Soc.*, v., 237.
   Hab. Ceylon.[1]

32. S. CONSANGUINEUS, Sauss.
   Priconemis consanguineus, Saussure, *Hym. Novara Reise*, 62.[1]
   Hab. Ceylon.[1]

34. S. COTESI, Cam., *postea*.
   Hab. S. India.

35. S. CRINITUS, Bing.
   Priconemis crinitus, Bingham, *Jour. Bomb. Soc. Nat. Hist.*, v., 238.[1]
   Hab. Ceylon.[1]

36. S. FULGIDIPENNIS, Sauss.
   Priconemis fulgidipennis Saussure, *Hym. Novara. Reise*, 61.
   Hab. Ceylon.[1]

37. S. GIGAS, Tasch.
   Priconemis gigas, Taschenberg, *Zeits. Gess, Naturwiss.* XXXIV., 40.[1]
   Hab. Java.

38. S. HUMBERTIANUS, Saus.
   Priconemis humbertianus, Saussure, *Hym. Novara Reise*.
   Hab. Ceylon.[1]

39. S. JUNO, Cam., *Infra*.
   Hab. Barrackpore.

40. S. MADRASPATANUS, Smith.
   Pompilus Madraspatanus, Smith, *Cat. Hym.*, III. 144, 130.[1]
   Hab. Madras,[1] Nicobar Islands.

41. S. MELLERBORGI, Dbm.
   Priconemis Mellerborgi, Dbm., *Hym. Eur.* I., 458.
   Hab. Java.[1]

42. S. MIRANDA, Cam. *postea*.
   Hab. Barrackpore.

43. S. PEDESTRIS, Sm.
   Pompilus pedestris, Smith, *Cat. Hym.*, III., 147, 139.
   Hab. Sumatra.[1]

44. S. PERPLEXUS, Sm.
   Pompilus perplexus, Smith, *Cat. Hym. Ins.* III., 147, 140.
   Hab. Sumatra.[1]

45. S. PEDUNCULATUS, Sm.
   Pompilus pedunculatus, Smith, *Cat. Hym. Ins.*, III., 145, 131.[1]
   India.[1]

46. S. PEREGRINUS, Sm. (Pl. III. f. 19.)
   Priconemis peregrinus, Smith, *Trans. Ent. Soc.*, 1875.
   Hab. Barrackpore (*Rothney*).

47. S. ROTHNEYI, Cam., *Infra*.
   Hab. Barrackpore.

48. S. SERICOSOMA, Sm.
   Pompilus sericosoma, Smith, *Cat. Hym.*, III., 146, 137.[1]
   Hab. Sumatra.[1]

49. S. OPTIMUS, Sm.
   Priconemis optimis, Smith, *Jour. Linn. Soc.*, II., 93, 5.[1]
   Hab. Singapore.[1]

50. S. VERTICALIS, Sm.
   Priconemis verticalis, Smith, *Proc. Linn. Soc.*, II., 94, 6.
   Hab. Borneo, Malacca.[1]

51. S. WAHLBERGI, Dbm.
   Priconemis Wahlbergi, Dbm, *Hym. Eur.*, I., 458.[1]
   Hab. Java.[1]

SALIUS ROTHNEYI, *sp. nov.*

Black, pruinose; eyes a little diverging beneath, straight. Ocelli in a triangle, separated from the eyes by a somewhat greater distance than they are from each other. Clypeus a little projecting towards the apex, which is broadly rounded and margined. Occiput transverse, rounded at the sides. There are two broad ridges above the antennæ, having a smooth, shining fovea between them. Head opaque, finely and closely punctured; the clypeus has the punctures distinctly separated, and it is more shining. The cheeks and clypeus bear a close silvery pubescence and some long silvery hairs; the vertex bears long fuscous hairs. Mandibles finely punctured at the base; the apex piceous. Pronotum short, in front obliquely transverse laterally; behind arcuate. Median segment shorter than the mesothorax, broadly and gradually rounded; thorax opaque, finely and closely punctured; median segment towards the apex transversely striolate; the central furrow indistinct. Abdomen shining, indistinctly pruinose; the apical segments finely punctured and covered with longish fuscous hairs; the terminal segment acute at apex; the petiole at the base about one third of the width of the apex. Antennæ shortish, pruinose. Wings hyaline, with a slight fuscous tinge; a cloud at the basal nervure; a broader one extending from the apex of the stigma to near the apex of the radial cellule. Second cubital cellule obliquely quadrate; at the top one-half longer than the third at the top; at the bottom a little shorter. The first recurrent nervure is received in about the apical third of the cellule; the second distinctly before the middle. Legs pruinose; the long spur of the hind tibiæ reaching to the middle of the metatarsus; the tibial spines stout, the central very thick, somewhat triangular at the apex, the metatarsal brush long, thick, the tarsal joints spined and pilose beneath; the front tarsi without a brush.

Length, 10 mm.

SALIUS COTESI, *sp. nov.* (Pl. III. f. 3).

Similar in the colouration of the body to *S. rothneyi*, as also in having in the wings three clouds, but abundantly distinct in structure. The clypeus at the apex is shining, and transverse in the middle; the elongated ridges above the antennæ, so prominent in *Rothneyi*, are absent, as is also the shining fovea, but there is a small carina there; the eyes distinctly diverge beneath; the ocelli are in a triangle, and closer to each other; the hinder being separated from the eyes by twice the distance they are from each other; the median segment at the apex is more abrupt; the abdomen is longer, being as long as the head and thorax united, and its apical segments are not so thickly haired; the form of the second cubital cellule is very different; the first transverse cubital nervure is elbowed at the middle, and bends towards the second, making the top of the cellule there about one-fourth of what it is at the bottom, and about one-third of the length of the top of the third; at the bottom, the second cubital cellule is about three-fourths of the length of the third; the second transverse cubital nervure is sharply elbowed at the top, making the cellule much narrower at the top than at the bottom, where it is rounded broadly at the apex, instead of acutely angled as in *Rothneyi*, while the cubital nervure terminates completely there; the first recurrent nervure is received a very little beyond the middle; the second at a less distance from the transverse cubital nervure than is the first; the radial nervure becomes elbowed about the basal third (and also more sharply), while in *Rothneyi* it turns up at the middle of the cellule. *S. cotesi* also is larger, being 13 mm. in length. The long spur of the hind tibiæ does not reach the middle of metatarsus; the tibial spines are shorter and fewer.

SALIUS (PRICONEMIS) PEREGRINUS, *Sm.* (Pl. III. f. 4).

Eyes curved above, slightly converging beneath. Ocelli almost in a triangle, the posterior separated from the eyes by a somewhat greater distance than they are from each other. Clypeus convex, transverse at the apex, the sides rounded. Prothorax shorter than the head; the sides obliquely dilated towards the head, convex above, the centre furrowed, the dilated part narrowing towards the base of the furrow; behind concave. Median segment at apex transverse, the apical part obliquely sloped, depressed in middle, and striolate. Radial cellule lanceolate at apex, elongate, narrow; third cubital cellule at top shorter, at bottom longer than the second. The second recurrent nervure is received shortly before the middle. In the hind wings the anal nervure is interstitial. The long spur of the tibiæ hardly reaches to the middle of the metatarsus. In the $\male$ the antennæ are nearly as long as the body, stout, tapering towards the apex; the third and fourth joints subequal. In the $\male$ the apical abdominal segment bears a thick longish tuft of black hairs.

SALUIS PEDUNCULATUS, *Sm.*

I have a number of specimens which are probably referrible to this species. Head alutaceous, the eyes curved, a little converging beneath; ocelli in pits, small, the hinder separated from the eyes by a greater distance than they are from each other. Head almost transverse in front, a little convex behind. Clypeus piceous and transverse in the middle at the apex; the sides rounded; mandibles yellow at the base. Pronotum slightly shorter than the head; slightly narrowed towards the head; rounded at the base; the apex a little curved inwardly. Median segment gradually rounded to the apex, as long as the mesothorax, transversely striated. Abdomen subpetiolate, the petiole becoming gradually dilated to the width of the second segment; the apical segments smooth, glabrous, except at extreme apex. The long spur of the hind tibiæ does not reach to the middle of the metatarsus, tibiæ sparsely spined. The second cubital cellule at top and bottom distinctly longer than the third; the first recurrent nervure is received a little beyond the middle; the second a little beyond the basal third. In the hind wings the anal nervure is received before the termination of the cubital. In length my specimens average 12 mm.

Saluis juno.

Black; the abdomen reddish, black at the base; the knees and fore tarsi rufo-testaceous; wings subhyaline, the apex smoky. Eyes curved, converging a little at the bottom. Ocelli separated from the eyes by nearly the same distance that they are from each other. Clypeus broadly convex; the sides at the apex oblique, the middle rounded. Occiput slightly concave. Prothorax nearly as long as the head, not much, if at all, narrowed towards the base. Median segment as long as the mesothorax, gradually rounded to the apex, irregularly transversely striated. Head and thorax alutaceous, covered with a pale pile. Abdomen pruinose; the petiole with a distinct neck at the base. Antennæ moderately elongate, microscopically pilose. The second cubital cellule at top and bottom considerably longer than the third; the first and third transverse cubital nervures obliquely curved; the second straight, slightly oblique; both the recurrent nervures received a little beyond the middle. The wings are rather short. Legs elongate, moderately stout, pruinose, the coxæ white with a silvery pile, the tibial spurs sparse, golden-fulvous; the apex of the hind tibiæ and the base of the tarsus bearing a thick fulvous pile; the long spur of the hind tibiæ does not quite reach to the middle of the metatarsus.

Length, 8 mm.

Hab. Barrackpore (*Rothney*).

POMPILUS.

Pompilus, Fab., *Ent. Syst. Supp.* 246.
Ferreola, Smith, *Cat. Hym.* III., 167.

1. POMPILUS ANALIS, Fab.
   Pompilus analis, Fabricius, *Syst. Piez.*, 188, 4; Dbm., *Hym. Eur.*, I., 47; Lep., *Nat. Hist. d. Ins. Hym.*, III., 439, 35.
   Hab. Common and widely distributed over our region, Singapore, Java, Bachian, Celebes, Aru.

2. POMPILUS ARIADNE, Cam., *postea*.
   Hab. Barrackpore.

3. P. BEATUS, Cam., *postea*.
   Hab. Bungalore.

4. POMPILUS BRACATUS, Bingh., *Jour. Bomb. Nat. Hist. Soc.*, V., 236.
   Hab. Pegu Hills.

5. POMPILUS BUDDHA, Cam, *infra*.
   Hab. Poona (*Wroughton.*)

6. P. CIRCE, Cam. Pl. III. f. 5.
   Ferreola fenestrata, Smith, *Cat. Hym. Ins.* III., 169[1].
   non Smith, *l.c.* p. 144.
   Hab. Madras[1], Poona (*Wroughton*).

7. P. COMPTUS, Lep.
   Nat. Hist. Ins. Hym. III., 425, 3[1].
   Hab. India[1].

8. P. COTESI, Cam. *infra*.
   Hab.

9. P. CORIARIUS, Tasch.
   Zeit. f. gess. Natur. Wessen. XXXIV., 49, 1.
   Hab. Java, Singapore[1].

10. P. DEHLIENSIS, Cam., *infra*.
    Hab. Dehli (*Rothney*).

11. P. DETECTUS, Cam.
    Hab. Barrackpore (*Rothney*).

12. P. DORSALIS, Lep.
    Nat. Hist. d. Ins. Hym., III., 407, 13.[1]
    Hab. India.[1]

13. P. DIMIDIATIPENNIS, Sauss.
    Ferreola dimidiatipennis, Saussure, *Hym. Novara Reise*, 47.[1]
    Hab. Ceylon.[1]

14. P. ELECTUS, Cam. *postea*.
    Hab. Barrackpore (*Rothney*).

15. P. FENESTRATUS, Sm.
    Cat. Hym. Ins., III., 144, 128.[1]
    Hab. Bengal.[1]

16. P. FASCIATUS, Bingh.
    Ferreola fasciata Journ., Bomb., Nat. Hist. Soc. V., 241, 12.
    Hab. Burmah.

17. P. GRAPHICUS, Sm.
    Cat. Hym. Ins., III., 148, 143.[1]
    Hab. Phillipines.[1]

18. P. GREENII, Bingh.
    Ferreola Greenii, Bingham, *Jour. Bomb. Nat. Hist. Soc.*, V., 240; 11.[1]
    Hab. Ceylon.[1]

19. P. HECATE, Cam., *postea*.
    Hab. Barrackpore (*Rothney*).

20. P. HONESTUS, Sm.,
    Smith, *Cat. Hym. Ins.* III., 144, 129.[1]
    Hab. India.

21. P. HERO, Cam. *postea.*
    Hab. Barrackpore.

22. P. IGNOBILIS, Saussure.
    Hymen. d. Novara Reise, 60.[1]
    Hab. Ceylon.[1] Sikim.

23. P. INCOGNITUS, Cam. *postea.*
    Hab. Barrackpore (*Rothney*).

24. P. LASCIVUS, Cam. *postea.*
    Hab. Barrackpore (*Rothney*).

25. P. LEUCOPHÆUS, Sm.
    Proc. Linn. Soc. II., 921.[1]
    Hab. Malacca.[1]

26. P. LUCIDULUS, Sauss.
    Homonotus lucidulus, Saussure, *Hym. d. Novara Reise,* 50.
    Hab. Ceylon.[1]

27. P. MACULIPES, Sm. (Pl. III, f. 16).
    Trans. Linn. Soc. VII., 186 1.[1]
    Hab. Manipuri, North West Provinces.

28. P. MIRANDA, Sauss.
    Ferreola miranda, Saussure, *Hym. d. Novara Reise,* 49.[1]
    Hab. Ceylon, Trincomalia.[1]

29. P. PARTHENOPE, Cam., *infra.*
    Hab. South East Provinces.

30. P. PEDALIS, Cam., *infra.*
    Hab. Barrackpore (*Rothney*).

31. P. PULVEROSUS, Smith.
    Proc. Linn. Soc. II., 93, 3[1].
    Hab. Borneo.[1]

32. P. ROTHNEYI, Cam. *postea.*
    Hab. Barrackpore (*Rothney*).

33. P. RUFO-UNGUICULATUS, Tasch.
    Zeits f. ges. Natur. Wissen XXXIV., 5, 4, 9.
    Hab. Java.

34. P. UNIFASCIATUS, Smith.
    Proc. Linn. Soc. III., 145. 133.[1]
    Hab. India,[1] Sumatra.[1]

35. P. VAGABUNDUS, Sm. (Pl. III. f. 23).
    Proc. Linn. Soc. II., 92, 2.[1]
    Hab. Borneo,[1] Barrackpore, Mussoure, (*Rothney*).

36. P. TRICOLOR, Sauss.
    Ferreola tricolor, Saussure, *Hymen. d. Novara Reise*, 48.[1]
    Hab. Singapore.[1]

37. P. VISCHNU, Cam., *postea.*
    Hab. Barrackpore.

38. P. VIVAX, Cam., *postea.*
    Hab. Barrackpore.

39. P. WROUGHTONI, Cam., *postea.*
    Hab. Poona (*Wroughton*).

40. P. ZEBRA, Cam., *postea.*
    Hab. Shellong.

41. P. ZEUS, Cam., *postea.*
    Hab. Barrackpore (*Rothney*).

Section—FERROLA.

FERROLA FENESTRATA, *Bingham.*

This is a distinct species from *fenestrata*, Smith, which has only the prothorax reddish. It is probably undescribed.
Hab. Burmah.

POMPILUS CIRCE, *Cam.* (Pl. III f. 5).

This is the most conspicuous species of the section. The collar is more elongated, is transverse at the apex in the middle, but curves round to the tegulæ at the sides; the clypeus is rounded; the ocelli small, in a curve, and separated from each other by a much greater distance than they are from the eyes.

POMPILUS PEDALIS, *sp. nov.*

Black, the basal two segments entirely, and the basal two-thirds of the third, red; the head and thorax densely covered with grey pile; the wings fusco-violaceous, the base to the transverse basal nervure subhyaline. Eyes arcuate, distinctly converging beneath. Ocelli large, in a curve, separated from each other by a much greater distance than they are from the eyes; the anterior in a pit; and an oblique short furrow runs from the posterior. Clypeus short, subarcuate. The head almost hoary with a greyish-white pubescence; on the top it is shorter, convex in front, concave behind. Occiput convex. Prothorax longer than the head, longer than broad, narrowed towards the head; at apex angled in the centre. Median segment as long as the prothorax; with a very slight slope above, the sides at the apex projecting into a longish sharp triangular tooth. Abdomen sessile, longer than the head and thorax united; pruinose, the apical segment impunctate. Antennæ short, about as long as the thorax, stout. Legs densely pruinose; the hinder tibiæ sparsely spined; the hind tibiæ not much longer than the metatarsus; the long spur of the hind tibiæ reaches to the middle of the latter. For wings see fig 6, pl. III. Claws bifid at apex; the tarsi without a brush.

This species differs from the other species here noticed, in the eyes being more arcuate at the top and converging much more at the bottom.

POMPILUS (FERREOLA) ARIADNE, *sp. nov.* (Pl. III. f. 7, 7*a*).

Black, the spurs white, palpi yellow, mandibles reddish, wings subhyaline. Head smooth and shining, sparsely pubescent; eyes arcuate, equally converging at top and bottom; ocelli large, in a curve; the posterior separated from each other by more than twice the distance they are from the eyes. Head convex in front, concave behind; antennæ placed immediately over the clypeus, over which the front projects; clypeus rounded at the apex. Prothorax quadrate, longer than the head, not narrowed towards the head; behind almost transverse. Pleuræ compressed, impunctate, almost glabrous. Median part of scutellum broad, a little narrowed towards the apex; median segment longer than the mesothorax; depressed in the middle at the apex; the lateral projections acutely triangular; the apex bearing depressed longish hairs. Abdomen sessile, compressed laterally; the third and following segments covered with dense silvery hairs. Legs stout; the tarsi testaceous; the long and stout calcaria longer than the metatarsus; claws bifid. For wings see fig. 7, pl. III.

Length, 6 mm.

Hab. Barrackpore (*Rothney*).

A very distinct little species.

POMPILUS (FERREOLA) HECATE, *sp. nov.* (Pl. III. f. 8).

Black, pruinose, the face densely covered with a silvery pile, the wings hyaline, the apex infuscated. Eyes broadly arcuate, a little converging beneath; ocelli large, almost forming a triangle, the posterior separated from each other by a greater distance than they are from the eyes. Clypeus short, transverse, the sides rounded. Head behind very little developed, and almost transverse. Prothorax not much longer than the head; almost transverse behind, not much narrowed in front. Median part of scutellum narrowed distinctly toward the apex; median segment longer than the mesothorax, depressed in the middle at apex; the laterally produced angles broad, short. Abdominal segments with a broad belt of silvery pruinose pubescence. Legs moderately long; the hinder tibiæ with few spines—longish and black. The transverse basal nervure is not interstitial; for neuration, see pl. III. f. 8.

Length, 7 mm.

Hab. Barrackpore.

From *P. Wroughtoni* which it resembles in colouration it may be known by being stouter, by the head being longer, by the eyes being nearer each other at the top; by the pronotum not having such a gradual slope to the head, and almost transverse behind, by the abdomen being shorter and broader and stouter.

POMPILUS (FERROLA?) ROTHNEYI, *sp. nov.* (Pl. III. f. 9).

Black, pruinose with a plumberous hue, the apex of the abdominal segments broadly black, not pruinose, the pruinosity giving the insect a greyish hue; wings yellowish hyaline, the apex infuscated. Eyes a little converging beneath, ocelli moderate, not in a triangle; the posterior separated from each other by a somewhat greater distance than they are from the eyes. Clypeus equally convex all over, short, broad, the sides obliquely truncated, the apex almost transverse; labrum half the length of the clypeus, bluntly rounded. Occiput transverse, very little developed behind the eyes. Antennæ moderately stout, the scape greyish pruinose. Prothorax longer than the head, not much narrowed towards the head, the sides a little convex, at apex arcuate. Median segment concave at the apex; the sides terminating in stout, somewhat triangular projections, the apex with a thick fringe of pale hair. Abdomen subsessile. Radial cellule short, broad in the middle, the basal abscissa of the radius a little longer than the second, which is straight, oblique and not curved. The second cubital cellule at the top more than twice the length of the third; at bottom not much longer than it; the third cellule very much narrowed at the top; the transverse cubital nervures being almost united; the first recurrent nervure received quite close to the apex of the wing; the second a little before the middle; it is elbowed in the middle. Legs densely pruinose; the spines long; the long spur of the hind tibiæ reaching before the middle of the metatarsus: the claw with a blunt, thick subapical tooth.

This species forms a transition to *Ferreola*, the apex of the median segment being only moderately concave and hardly dilated at the sides; the antennae, too, are higher up over the clypeus, and the anal nervures in the hind wings are received beyond the cubital. In one example the third cubital cellule is distinctly petiolated.

Length 12 mm.

P. Wroughtoni has the apex of the median segment more as in the typical *Ferreola*, *i.e.*, it is produced laterally, but not quite so much as in, say, *Circe*, it forming, in fact, a regular curve, and it is also depressed in the middle; the abdomen is compressed, the anal nervure in the hind wing is interstitial; the antennæ are placed immediately over the clypeus; the head is very little developed behind the eyes; the basal nervure is not interstitial; the centre of scutellum not much, if any, narrowed towards the apex— the pubescence on the edge of the pronotum forms a whitish band.

POMPILUS WROUGHTONI, *sp. nov.* (Pl. III. f. 10).

Very similar to *P. Rothneyi*, having the same grey pruinose vesture, with the abdominal segments grey and black; and the apex of the median segment concave, the head very little developed behind the eyes and the abdomen subsessile; but is smaller, narrower, and more slender ; the wings are subhyaline throughout, not yellowish or infuscated at the apex ; the second cubital cellule is much longer at the bottom compared with the third ; the third being of the length of the space bounded by the first transverse cubital and the first recurrent, the latter being received at a greater distance from the transverse cubital ; the second recurrent is received in the apical fourth of the cellule, not before the middle, and lastly the long spur of the metatarsus reaches almost close to the apex of the metatarsus.

— POMPILUS DELHIENSIS, *sp. nov.* (Pl. III. f. 11).

Black, densely covered with a silvery pubescence, especially thick on the face, median segment and on the apices of the abdominal segments ; wings yellow, a broad fuscous band at the radial cellule. Head slightly convex in front, more deeply concave behind. Eyes slightly arcuate at top, at bottom almost parallel ; ocelli large, forming almost a triangle ; the hinder separated from each other by a greater distance than they are from the eyes. The front with an obscure furrow. Clypeus rounded bluntly and rufous at apex ; the centre with a minute incision ; mandibles ferruginous, black at top ; palpi fuscous. Prothorax a little shorter than the head, the sides slightly convex. Median part of scutellum not much narrowed towards the apex. Median segment shorter than the mesothorax, above with a gentle slope ; the apex oblique, with a slight inward curve. Abdomen subsessile ; curved, a little longer than the head and thorax united ; the segments with a silvery band at the apices: the apical segment acute, shining, impunctate, and bearing a few long blackish hairs. Legs stout, densely pruinose, the tibiæ and tarsi thickly spined ; the base of hind tibiæ with a white mark behind. The spurs white, reaching to the middle of the apex ; claws with a narrow subapical tooth. The basal nervure interstitial ; the anal nervure in hind wings received before the termination of cubital. (For neuration, see pl III. f. 11.)

Length 9 mm.

Hab. Delhi, (*Rothney*).

Is very nearly related to *P. Rothneyi*, which it also resembles in having the apex of the median segment sub-concave ; but differs in the wings having the apex hyaline, the cloud not extending to it ; the third cubital cellule is much wider at the top, in the radial cellule being longer and much narrower ; in the prothorax having the sides convex, in its apex being transverse, not arcuate ; in the scutellum being shorter, broader, and not much narrowed towards the apex, and in the white spurs and base of hinder tibiæ, the tarsi, too, being much more thickly spined and fringed beneath.

POMPILUS HERO, *sp. nov.* (Pl. III. f. 12).

Black, densely pruinose, a white belt of pubescence on the pronotum and on the abdominal segments, the metanotum, apex of median segment and base of abdomen tufted with thick greyish hair ; the scape yellow beneath ; the flagellum ferruginous ; the 2-4 tarsal joints white, black at the apex ; the edge of pronotum and tegulæ yellowish, wings yellowish-hyaline, infuscated at the apex. Head distinctly convex in front, indistinctly so behind ; the clypeus covered with a silvery pubescence, the rest with a short pile and with longish soft pale hairs. Clypeus arcuate in the centre, the sides obliquely truncated. Eyes slightly arcuate at the top, converging a little at the bottom ; ocelli separated from each other by a greater distance than they are from the eyes, not forming a triangle. Prothorax shorter than the head, gradually narrowed towards it, behind arcuate. Scutellum gradually narrowed towards the apex. Median segment shorter than the mesothorax, with a gentle slope, the apex oblique. Abdomen sessile; the apical segment citron-yellow, densely covered with a pale pubescence, and at the apex with longish black hairs. Legs stout; the tibiæ and tarsi with few spines; the claws bifid, the shorter claw much thicker than the other. Antennæ longish, stout, the apical joints dilated beneath. Basal nervure interstitial; the anal in hind wing being received beyond the cubital. (For neuration see pl. III. f. 12). Length 11 mm. ♂. Claws with the basal tooth stout, not reaching to the apex.

In one specimen the hinder tibiæ are whitish-yellow at the base, the spurs being also of this colour.

POMPILUS INCOGNITUS, *sp. nov.* (Pl. III. f. 13).

This species agree in the colouration of the body, legs, and wings with *P. pedestris*, Smith — having the body densely cinereous pruinose, the hind femora and tibiæ red, the wings fusco-hyaline, deeply infuscated at the apex, and the abdomen with cinereous bands; but it must be, I should think, distinct, *e.g.*, although the apex of the median segment is truncated, yet it can hardly be said to be "produced laterally, forming obtuse tubercles"; and the third cubital cellule is called "subtriangular," while here it is distinctly petiolated and not sub-triangular.

Eyes distinctly converging beneath; ocelli separated from the eyes by about the same distance they are from each other. Clypeus a little convex, short, broad; the apex transverse. Head very little developed behind the eyes: the occiput a little concave. Prothorax a little longer than the head, having a gradually rounded slope towards the head, and sub-quadrate behind, arcuate, angled in the middle. Median segment with a slight slope to near the apex, when it becomes oblique; the apex transverse, bearing a thick silvery pubescence; the metapleuræ projecting sharply at the apex into tubercles [this may be the obtuse tubercles of Smith]. Abdomen elongate, narrow, sessile, longer than the head and thorax united; sharply pointed at the apex, and bearing some long black hairs; the apical segment very smooth and shining. Antennæ shorter than the abdomen, tapering perceptibly towards the apex, not convolute. Wings comparatively short; the radial cellule about twice longer than wide; the radial nervure curved at both ends; second cubital cellule at top half the length of the bottom, where it is a little longer than the third; the third with a petiole as long as three-fourths of the top of the second cubital cellule; the third cellule narrowed at the top, but not forming a triangle, both the nervures being distinctly curved; the first recurrent nervure very oblique and received near the apex of the cellule; the second in the middle. Legs pruinose; the spines long, black; the base of the hind femora and apex of the tibiæ black; the long spur of the hind tibiæ reaches to the middle of the metatarsus. The cloud in the fore wings commences at the apex of the radial cellule.

What is probably a variety has the hind tibiæ black; this form being also smaller.

Length, 12 mm.

POMPILUS VIVAX, *sp. nov.* (Pl. III. f. 14).

Black, pruinose, the scape beneath, the edge of the pronotum and tegulæ yellowish, the face, the scutellum, apex of median segment, coxæ, and base of abdomen densely covered with a thick greyish silvery pubescence; wings subhyaline, the apex infuscated; second cubital cellule petiolate. Eyes a little converging beneath; ocelli in a curve, the hinder separated from the eyes by a distinctly less distance than they are from each other; apex of clypeus in the middle forming a shallow curve; the sides oblique; occiput transverse; the sides rounded. The pubescence below the antennæ is very dense ; on the front sparser. Prothorax a little longer than the head ; the sides straight, a little narrowed in front, the apex acutely incised against the mesonotum. Median segment with a very gentle slope to the apex which is rounded. Abdomen sessile. Legs long, densely pruinose ; the tibial spine long ; the spurs pale, reaching to near the apex of the metatarsus ; claws bifid. Radial cellule not much longer than deep, narrowed rather sharply in the middle. The first transverse cubital nervure roundly elbowed in the middle, at top three-fourths of the length of the bottom ; second cubital cellule shortly pedunculated, subtriangular ; the first recurrent nervure received near the apex, the second a little beyond the middle. Antennæ stout, the apical joints dilated in the middle ; the scape yellow, joints three and four brownish beneath.

Length, 8 mm.

Hab. Barrackpore (*Rothney*).

POMPILUS VISCIINU, *sp. nov.*

Identical in the colour of the body and wings to *P. vivax* ; differing in the second cubital cellule not being petiolate ; the scutellum and apex of median segment not bearing a dense pubescence ; in the ocelli forming a triangle and the posterior being, if anything, separated from the eyes by a somewhat greater distance than they are from each other ; in the clypeus being rounded ; in the occiput being slightly convex ; in the median segment being widely and deeply furrowed down the centre, in the spurs being shorter (three-fourths of the length of the metatarsus) and black. The abdominal segments have a broad belt of greyish pruinose pubescence on the apex. The legs are pruinose ; the spines short, petiole moderately narrow at the base, becoming gradually wider towards the apex.

Length, 6 mm.

Pompilus unifasciatus, Sm.

A specimen in Mr. Rothney's collection is thus named by Smith. The type has the head entirely yellow; but this example has a broad black band on the vertex and front. Head convex in front, transverse, with the sides rounded behind. Ocelli almost in a triangle, the posterior separated from the eyes by about the same distance that they are from each other; eyes arcuate above, beneath parallel; clypeus transverse at the apex, the sides obliquely rounded. Prothorax shorter than the head, the sides rounded, narrowing towards the head; behind arcuated, bluntly angled in the centre; there is a furrow in the middle of the pronotum. Legs sparsely spinose; the spines long; the long spur of the hind tibiæ reaches beyond the middle of metatarsus. The second recurrent nervure received in the apical third of the cellule; the anal in hind wing received beyond the termination of the cubital.

Pompilus electus, sp. nov. (Pl. III. f. 15).

Black; the basal two and the greater part of the third segment red; the greater part of the front and the base of the four posterior tibiæ reddish; the tarsi inclining to fuscous; wings hyaline, a small band along the basal nervure, and a broad one extending from the base of the stigma to the third transverse cubital nervure, fusco-violaceous. Head as broad as the thorax; moderately convex in front, almost transverse behind. Eyes broadly arcuate above, almost diverging below; ocelli hardly forming a triangle, separated from each other by a slightly greater distance than they are from the eyes. Clypeus projecting a little; the sides oblique, straight, the apex bluntly rounded. The head closely punctured; the clypeus and cheeks densely covered with silvery pubescence, the frontal furrow indistinct. Antennæ longish, filiform.

Prothorax shorter than the head; the sides almost straight, not much narrowed towards the head. Median segment short, rather abruptly rounded towards the apex; there is a patch of white silvery hair on either side at the apex. The thorax can hardly be said to be aciculate, and has an olive tinge in parts. Abdomen longer than the head and thorax united, subpetiolate; the apical segment glabrous and impunctate at the base; the rest bearing long hairs and the apex a depressed rufous stiff pile. Legs longish, the hind tibiæ serrate; the long spur of the hind tibiæ reaches to a little beyond the basal third of the metatarsus; the legs densely covered with a silvery pile. Wings longer than the body; radial cellule elongate, narrow, lanceolate at base and apex; the second cubital cellule at top, about one-third longer, at bottom not much longer than it; the first recurrent nervure received beyond the middle, about the same distance that the second is received from the base of the cellule.

Length, 7—8 millim.

The serrated tibiæ are pretty much as in *Priconemis*; but the transverse basal nervure is interstitial, and there is no furrow at the base of abdomen beneath. The wings and antennæ are longer than in any other Indian species known to me from autopsy.

✝POMPILUS BUDDHA, *sp. nov.* (Pl. III. f. 20).

Black; the abdomen and legs red; the clypeus, inner orbits somewhat widely, and the outer narrowly; three lines on the pronotum (a large central and a somewhat shorter lateral) and two lines on it behind, yellow; wings hyaline, deeply infuscated from the base of the stigma to the apex, which is pale. Eyes almost parallel; ocelli separated from each other by a somewhat greater distance than they are from the eyes. Occiput transverse. Clypeus short, broad, projecting a little, the apex broadly rounded. Prothorax shorter than the head, the sides rounded. Head and thorax alutaceous, pruinose. Petiole without a neck, gradually enlarged towards the apex. Antennæ elongate, moderately stout. The radial cellule moderately wide, the apex sharply lanceolate; the first cubital cellule hardly twice longer than the second, the first transverse cubital nervure broadly curved, the second straight; the third elbowed sharply at the middle; the top of the cellule being thus much narrowed; the first recurrent nervure is received in the apical third; the second almost in the middle of the cellule. Legs elongate, pruinose, the tibial spines few; the long spur of the hind tibiæ reaches to the middle of the metatarsus. Claws with a short stout sub-basal tooth.

Length 7-8 mm.

POMPILUS ZEUS, *sp. nov.* (Pl. III. f. 21).

Black; the basal three abdominal segments, the hind femora and tibiæ, the middle femora, except at the base, red; the spines glistening white, wings fusco-hyaline, with fuscous nervures, tegulæ yellowish. Head a little wider than the thorax; eyes very slightly arcuate above, converging beneath; ocelli separated from each other by a distinctly greater distance than they are from the eyes. Clypeus gaping, the sides rounded, the apex almost transverse. Antennæ stout, brownish beneath, the third and fourth joints subequal. Prothorax scarcely so long as the head, the sides straight to the base, behind almost transverse. Median segment somewhat longer than the prothorax; the base with a very gradual slope, the apex much more abrupt; the surface hid by a short close white pubescence. Abdomen longer than the head and thorax united, subsessile; the apical segment impunctate. Legs densely covered with a silvery pile; moderate, the tibiæ and tarsi sparsely spined. The long spur of the hind tibiæ reaches beyond the middle of the metatarsus. Wings two-thirds of the length of the body. For nervures see Pl. III. fig. 21.

Length, 8 mm.

The third cubital cellule is shorter than in any other Indian species I have seen.

—POMPILUS BEATUS, *sp. nov.* (Pl. III. f. 22).

Black, the pronotum with a broad yellow band; the three basal segments of the abdomen, except the apex laterally of the third, red; wings fusco-violaceous. Head small, narrower than the thorax, convex in front, and to a less extent behind. Eyes sharply arcuate at the top, reaching well back behind laterally; converging a little below. Ocelli in a curve, separated from the eyes by a less distance than they are from each other. Head longish from the front view, the clypeus being produced below the eye; its apex transverse. Clypeus and cheeks densely covered with a dense silvery pubescence. A narrow furrow on the front. Prothorax a little longer than the head, broadly arcuate behind, narrowed a little towards the head. Median segment with a gradual slope, and with a transverse ridge at the apex. Abdomen sessile, very gradually and slightly narrowed towards the apex, pruinose; the two apical segments densely covered with silvery pubescence. Legs stout, the hinder tibiæ with the spines of moderate thickness and length; the long spur of the hind tibiæ reaching close to the apex of the metatarsus. Antennæ short, stout, tapering towards the apex. Second cubital cellule sub-petiolate. For neuration see pl. III. fig. 22.

Length, 12 mm.

Hab. Bangalore, South India (*Mus. Cal.*).

POMPILUS VAGABONDUS, *Sm.* (Pl. III. f. 23).

Eyes arcuate above, parallel, not converging beneath. Ocelli in a curve, separated from each other by a greater distance than they are from the eyes. Clypeus transverse, the sides rounded. Head slightly convex in front. Prothorax shorter than the head, rounded in front. Median segment short, gradually rounded to the apex, not furrowed, obscurely aciculated. Abdomen subsessile; the pygidium elongate, sharply pointed at the apex, longitudinally rugosely striolate. The first transverse cubital nervure slightly curved, oblique; the second and third straight, converging at the top; the second cubital cellule at top and bottom twice the length of the third; both recurrent nervures received towards the apical third of the cellules.

POMPILUS FENESTRATUS, *Smith* (Pl. III. f. 24).

Eyes arcuate, converging a little at the base. Ocelli in a curve, separated by about the same distance from each other that they are from the eyes. Clypeus short, broad, the sides rounded, the apex very slightly arcuate. Head in front convex; the occiput transverse. Prothorax as long as the head, the sides not convex. Median segment aciculated; broadly furrowed down the centre. Abdomen subsessile; pygidium coarsely rugose, covered with long, stiff black hairs. Radial cellule acute in the middle; the third cubital cellule narrowed to a point above, the transverse cubital nervures almost touching there. The first transverse cubital nervure broadly curved; the first recurrent nervure is received a little beyond the middle; the second about the middle.

POMPILUS DETECTUS, *sp. nov.* (Pl. III. f. 25).

Black, the basal two, and the greater part of the third abdominal segment, red; densely pruinose; the wings fusco-violaceous. Eyes arcuate above, slightly converging beneath. Eyes in a triangle, separated from each other by about the same distance they are from the eyes. Clypeus short, sub-arcuate at the apex. Occiput transverse. Clypeus and cheeks covered with a dense short whitish pile. Front and vertex obscurely alutaceous. Prothorax a little shorter than the head; and with a rounded slope to the head. Median segment with a gradually-rounded slope to the apex. Abdomen subsessile, as long as the head and thorax united; pruinose, the apical segment coarsely rugose, covered with long bristly stout hairs. Legs stout, the hinder tibiæ with five rows of long stout spines; the long spur of the hinder tibiæ reaches beyond the middle. (For wings see pl. III. fig. 25.) Claws with a short submedian tooth.

POMPILUS LASCIVUS, *sp. nov.* (Pl. III. f. 26).

Black; the head, prothorax, mesonotum, with scutellum and metanotum, red; the wings with the basal half hyaline, the apical fusco-violaceous, except the extreme apex. Head wider than the thorax; the eyes almost parallel; the ocelli hardly forming a triangle; separated from the eyes by a distinctly greater distance than the hinder are from each other. Clypeus convex, the apex rounded. Prothorax shorter than the head, arcuated behind. Median segment with a gradually rounded slope, longer than the prothorax, transversely striolate. Abdomen subsessile, as long as the head and thorax united; with an olive tint, pruinose; the apical segment shining, impunctate. Antennæ stout. Legs stout, the tibiæ sparsely spined; covered with a silvery pubescence; the long spur of the hind tibiæ reaches a little beyond the middle. (For wings see pl. III. f. 26.) The entire body is more or less pruinose; the head and thorax semi-opaque, coarsely aciculate.

Length, 7 mm.

POMPILUS ZEBRA, *sp. nov.* (Pl. III. f. 27).

Black, the mandibles, apex of clypeus, inner orbits of the eyes to near the top broadly, the outer narrowly, a broad band on the pronotum, tegulæ, the abdomen with a band on the base of the second segment, the third entirely on the others, except a band on the base of the fourth, the apex of the femora broadly, the tibiæ and tarsi and the antennæ dull ferruginous; the head and thorax bearing long white hairs. Head a little wider than the thorax; the eyes arcuate above, the rest parallel; ocelli in a triangle, separated from each other by about the same distance they are from the eyes. Clypeus short, rounded at the apex. Prothorax a little longer than the head, narrowed gradually towards the base. Median segment about as long as the prothorax, gradually rounded to the apex; the apical half bearing a dense covering of white hair. Abdomen semisessile, a little longer than the head and thorax united; its apex moderately acute; the apical segment aciculate. Legs densely pruinose, stout; the tibiæ with reddish spines, widely separated; the three middle being the longest; the long white spur of the hind tibiæ reaches beyond the middle of the metatarsus. Claws with a thick basal tooth. There is no apparent sculpture on the body; there is a narrow furrow in the centre of the front; the occiput convex. The stigma is obscure testaceous; the hind wings are only infuscated at the apex.

Length, 10—11 mm.

Hab. Shillong.

POMPILUS PARENTHOPE, sp. nov.

Black; the wings fusco-violaceous. Eyes almost parallel. Ocelli separated from the eyes by a distinctly greater distance than they are from each other. Clypeus with the sides rounded; the middle slightly waved and margined. Head moderately well developed behind the eyes; the occiput a little concave. Pronotum hardly so long as the head; the sides rounded. Median segments a little longer than the prothorax, having a gradually rounded slope to the apex; the middle with a wide shallow furrow; alutaceous, covered with a fulvous down. Abdomen shining; the petiole becoming gradually wider towards the apex, so that it is there more than twice the width of the base. Apical segment rugose, thickly covered with stiff hairs; the sides and lower surface with long pale soft hairs. The second cubital cellule at the top more than twice the length of the third; at the bottom equal in length to it; the third at the top about one-third of the length of the bottom; third transverse cubital nervure with a gradual curve to the top; the first recurrent nervure is received near the apex; the second a little beyond the middle. Legs pruinose; the spines sparse; the long joint of the hind tibiæ short, not reaching to the middle of the metatarsus.

Length, 15 mm.

Hab. South-East Provinces.

Planiceps and *Aporus*, distinguished from *Pompilus*, Sensu str., by having only two cubital cellules, are treated by Kohl as sections of Pompilus.

PLANICEPS ORIENTALIS, *sp. nov.* (Pl. III. f. 1).

Black, shining, pruinose; the wings fusco-violaceous, with subhyaline clouds. Clypeus at apex, subarcuate, short, the sides obliquely truncated. Clypeus and cheeks to the antennæ thickly covered with a pale silvery pubescence; the rest of the head, shining, impunctate, very sparsely pilose. Ocelli in a curve, the hinder separated from the eyes by a distinctly less distance than they are from each other; behind them is a longitudinal furrow. Occiput transverse. Prothorax longer than the head, arcuate behind, laterally slightly convex. Median segment short, the base with a moderately rounded slope; the apex oblique. On either side towards the base is a deep semicrescentic short furrow. Abdomen longer than the head and thorax united, subsessile, acutely pointed; the apical segment shining, impunctate, bearing a few hairs. (For wings see pl. III., fig. 1.) The hind wings subhyaline, smoky at apex. The legs are stout, the two hinder tibiæ and tarsi stoutly spined. Antennæ stout, short, the third joint about one-quarter longer than the fourth.

APORUS COTESI, *sp. nov.* (Pl. III. f. 2).

Black, the scape beneath, palpi, the abdomen, except the two apical segments and the femora beneath, reddish; the tegulæ yellow; wings subhyaline, deeply infuscated from the second transverse cubital nervures. Head transverse behind, the sides rounded; eyes straight, slightly converging beneath; ocelli in a triangle, separated from the eyes by a greater distance than they are from each other; clypeus bluntly rounded at the apex. Vertex and front in the centre shining, almost glabrous; the rest of the face covered with a dense silvery pubescence. Mandibles yellow, piceous at the apex. Occiput transverse. Prothorax nearly as long as the head, angulated in the middle, not semicircular at the apex. Median segments as long as the mesothorax laterally; broadly furrowed down the centre, and having a gradual slope to the apex. The entire thorax covered with a white thick pubescence. Abdomen a little longer than the head and thorax united, subsessile; impunctate, shining, sparsely covered with a pale pubescence. (For wings, see pl. III., f. 2.) Legs moderately long; the tibiæ and tarsi with few spines; the long spur of the hind tibiæ reaching to the middle of metatarsus.

APORUS BENGALENSIS, *sp. nov.*

Black; the head and thorax appearing plumbeous, through being pruinose; the base and apex of the first and the apex of the second with a broad pruinose band; wings subhyaline, infuscated from the second transverse cubital nervure. Head almost transverse in front, concave behind. Eyes arcuate at top, almost converging at bottom. Ocelli in a triangle, the hinder separated from each other by a distinctly greater space than they are from the eyes. Antennæ longish, brownish underneath. Clypeus convex, almost transverse at the apex in the middle; the sides obliquely rounded. Prothorax a little longer than the head; almost transverse behind; the pleuræ bulging out on the lower side; and excavated broadly behind this bulge; the sides at top straight, not much narrowed till near the head. Median segment longer than the mesothorax, with a gradual slope to the apex; towards the apex with some indistinct waved striæ. Abdomen subsessile. Legs moderately long, the tibiæ with a few spines; the tarsi without them except at the apices of the joints; the long spur of the hind tibiæ reaches a little beyond the middle of the metatarsus. Wings short.

Length, 6 mm.

Apart from colouration, this species is very distinct from *A. cotesi* ♀. The head is concave behind, the collar is longer than the head, the prothorax almost transverse behind; the medium segment distinctly longer than the mesothorax, and the wings are shorter; the apical segment is rugose, densely covered with a stiff reddish pile, and at the apex with some longish hairs. It is a true *Aporus*.

(*Received June 23rd, 1891.*)

To the above are to be added the following, mostly very inadequately, described species of Mr. F. Smith in his posthumous work, "New Species of *Hymenoptera* in the British Museum."

Pompilus clotho, p. 146. Sumatra.
    ,,   *lachesis*, p. 146. Sumatra.

Pompilus atropus, p. 146. Sumatra.
„ *familiaris*, p. 147. Sumatra.
„ *pruniosus*, p. 147. India.
„ *capitosus*, p. 147. Burma.
„ *mitis*, p. 148. Bombay District.
„ *ephippiatus*, p. 148. Bombay District.
„ *multifasciatus*, p. 148. Bombay.
„ *decoratus*, p. 149. Bombay.
„ *simillimus*, p. 149. Calcutta.
„ *elegans*, p. 150. India.

The following species has been omitted from the alphabetical list :—

HEMIPEPSIS? SYCOPHANTE.

Gribodo, *Ann. Mus. Genoa*, I., 359.
Hab. Burma.
One of the species belonging to *flava* group.

DOLICHURUS.

This genus is usually placed in the *Pompilidæ*, but there can be little doubt but that its true location is with *Ampulex* and *Rhinopsis*.

1. DOLICHURUS TAPROBANÆ.

Smith, *Trans. Ent. Soc.*, 1869, 304.
Hab. Ceylon.

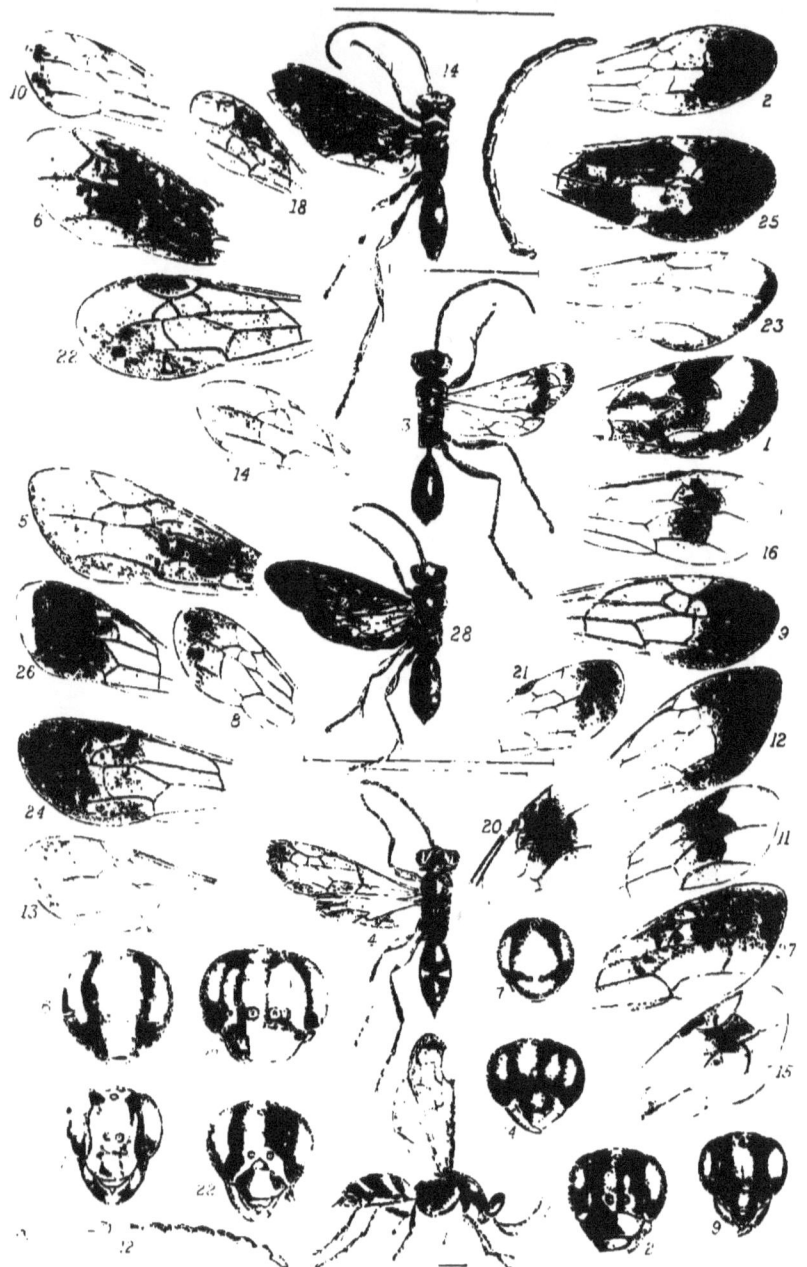

Explanation of Plate.

1. *Planiceps orientalis*, wing.
2. *Aporus cotesi*, wing.
3. *Pseudagenia festinata*, ♀.
4. *Ceropales orientalis*, ♀.
5. *Pompilus circe*, wing.
6. Do. *pedalis*, wing.
7. Do. *ariadne*, ♀.
8. Do. *hecate*, wing.
9. Do. *Rothneyi*, wing and head.
10. Do. *Wroughtoni*, wing.
11. Do. *Dehliensis*, wing.
12. Do. *hero*, wing and antenna.
13. Do. *incognitus*, wing.
14. Do. *vivax*, wing, head and antenna.
15. Do. *electus*, wing.
16. Do. *maculipes*, wing.
17. Do. *leucophæus*, head.
18. Do. do. wing.
19. *Salius peregrinus*, ♀.
20. *Pompilus Buddha*, wing.
21. Do. *zeus*, wing.
22. Do. *beatus*, wing and head.
23. Do. *vagabundus*, wing.
24. Do. *fenestratus*, wing.
25. Do. *detectus*, wing.
26. Do. *lascivus*, wing.
27. Do. *zebra*, head.
28. *Salius fulvipennis*, ♀.

From the Fourth Volume of the Fourth Series of "MEMOIRS AND PROCEEDINGS OF THE MANCHESTER LITERARY AND PHILOSOPHICAL SOCIETY."
Session 1891-92.]

Hymenoptera Orientalis; or Contributions to a knowledge of the Hymenoptera of the Indian Zoological Region. IV.

BY

P. CAMERON.

MANCHESTER
36, GEORGE STREET.
1892

Hymenoptera Orientalis; or Contributions to a knowledge of the Hymenoptera of the Indian Zoological Region. By P. Cameron. Communicated by John Boyd.

(*Received December 29th, 1891.*)

PART IV.

SCOLIIDÆ.

The identification of these striking insects is rendered very easy by the well-known *Catalogus Specierum Generis Scolia* of Messrs. Saussure and Sichel. I have followed their arrangement of the species, and have adopted the genera as defined by them, namely:—

Liacos, with its two divisions of *Triliacos* and *Diliacos*;
Scolia „ „ „ „ „ *Triscolia* and *Discolia*;
Elis „ „ „ „ „ *Trielis* and *Dielis*.

LIACOS.

I. *With three closed cubital cellules* = Triliacos.

1. LIACOS ANALIS.

Scolia analis, Fab., *Syst. Piez.*, p. 245, 37.

Scolia dimidiata, Guerin, *Voy. de Coq.*, II., p. 247; Burmeister, Scolidæ, p. 15, 2; Smith, *Cat. Hym.* III., 138; *Jour. Linn. Soc.*, IV., p. 118, 10.

Campsomeris Urvillii, Lep., *Hym.* III., 503, 12; Smith, *Cat. Hym.*, III., p. 112, 127.

Scolia Penangensis, Saussure, *Mélanges Hym.* p. 39, 17, ♀, var. b

Liacos (Triliacos) analis, Saussure and Sichel, *Cat. Specierum gen. Scolia,* p. 33.

Hab. Java, Borneo, Malacca, Philippines, Mollucas, Celebes Bouru, Sulu, Senegal.

— 2. Liacos fulvo picta, *sp. nov.*

Black, the third and following segments, pale fulvous, and covered with long fulvous hairs; the wings deep violaceous. Head and thorax marked with large, clearly separated punctures, the punctures on the head and pleuræ smaller than on the mesonotum; the front above and between the antennæ raised; the apex slightly incised; the surface longitudinally punctured; and above it is an impunctate border, which extends to near the middle of the eye incision. The front ocellus in a pit. The mesonotum has the furrows slightly curved, and extending from the apex to a little beyond the middle; the apex of the scutellum is impunctate, and has a short longitudinal channel in the centre at the apex. The median segment shallowly concave at the apex; the sides convex; there are two converging furrows in the middle; the space on the outer side of these at the base being impunctate. Abdomen punctured like the thorax; the hairs on the basal two segments black; and they have a bluish tinge; the base of the third segment is black, the black being dilated in the middle. The antennæ bare, dull black; the joints not dilated; the third joint distinctly shorter than the fourth. The second cubital cellule is dilated at the top; the transverse cubital nervures bulging out.

Length, 25 mm.

Allied to *analis* and *erythrosoma,* but easily known by the fulvous, not red, apex of abdomen; differing otherwise in the head being much more strongly punctured, this being also the case with the thorax.

Hab. Barrackpore (*Rothney*).

3. LIACOS ERYTHROSOMA.

Scolia erythrosoma, Burmeister, *Mon. Scol.* 15[1]; Smith, *Cat. Hym.*, III., 113, 134.

Scolia dimidiata, Smith, *l.c.*; *Journ. Linn., Soc.* VII., 29 12, part, var. from Bachian and Gilolo.[2]

Liacos (Triliacos) erythrosoma, Sauss. and Sichel, *Cat. Scol.* 35.

Hab. Poona, Sumatra[1]; Bachian and Gilolo.[2]

Smith, *Jour. Linn. Soc.*, 1869, p. 348, sinks *erythrosoma* as a variety of *analis*.

II. *With two closed cubital cellules* = Diliacos.

4. DILIACOS SICHELI.

Liacos Sicheli, Saussure, *Stett. Ent. Zeit.*, 1859, p. 172, f. 1, ♀; *Cat. Scol.*, 36.[1]

Hab. Sumatra.[1]

SCOLIA.

I. *Three cubital cellules* = Triscolia (Species 1—16).

1. SCOLIA NUDATA.

Scolia nudata, Smith, *Cat. Hym.*, 110, 120; Sauss. and Sichel, *Cat. Scol.*, 38, 7.[1]

Hab. North Bengal.[1]

2. SCOLIA BREVICORNIS.

Scolia brevicornis, Saussure, *Stett. Ent. Zeit.*, 1858, p. 198, 2; Sauss. and Sichel, *Cat. Scol.*, p. 39, 8[1].

Hab. Java,[1] Borneo.[1]

3. SCOLIA KOLLARI.

Scolia kollari, Saussure, *Stett. Ent. Zeit.*, 1858, p. 174; Sauss. et Sichel, *Cat. Scol.*, 40[1].

Hab. Java[1].

4. SCOLIA FORAMINATA.

Scolia foraminata, Saussure, *Stett. Ent. Zeit.*, 1858, p. 173; Sauss. et Sichel, *Cat. Scol.*, p. 40.[1]

Hab. Java.[1]

5. SCOLIA UNIMACULATA.

Scolia unimaculata, Kirby, *Trans. Ent. Soc.*, 1889, 446.[1]

Hab. India.[1]

6. SCOLIA TYRIANTHINA.

Scolia tyrianthina, Kirby, *l.c.* pl. XV., fig. 2.[1]

Hab. Andaman Islands.[1]

7. SCOLIA VELUTINA.

Scolia velutina, Saussure, *Stett. Ent. Zeit.*, 1858, 175; Sauss. et Sichel, *Cat. Scol.* 41, 13[1].

Hab. Java.[1]

8. SCOLIA OPALINA.

Scolia opalina, Smith, *Proc. Linn. Soc.* II., 89, 9[1]; Sauss. et Sichel, *Cat. Scol.* 45.

Hab. Borneo.[1]

9. SCOLIA PROCER.

Scolia procer, Illiger, *Mag.*, I., 196, 26; Fab., *Syst. Piez.*, 238; Burmeister, *Mon. Scol.*, 19, 9; Lepel, *Nat. Hist. Hym.*, III., 519, 3; Sauss. et Sichel, 43, 16.[1]

Scolia capitata, Fab., *Syst. Piez.*, 239, 3; Smith, *Cat.*[1] *Hym.*,[2] III., 111, 122.

Scolia patricialis, Burm., *Mon. Scol.*, 19, 10.

Hab. Java,[2] Sumatra,[2] Singapore,[2] Borneo,[1] Moluccas, Malacca.[1]

Smith, *Journ. Linn. Soc.*, 1869, p. 343, regards *patricialis* as a good species, but does not indicate the essential points in which it differs from *procer*.

10. SCOLIA SPECIOSA.

Scolia speciosa, Smith, *Proc. Linn. Soc.*, II., 90, 10[1]; Sauss. et Sichel, *Cat. Scol.* 44, 17.[1]

Hab. Borneo.[1]

11. SCOLIA MAGNIFICA.

Scolia magnifica, Sauss., *Stett. Ent. Zeit.*, 1859, 175; Sauss. and Sichel, *Cat. Scol.*, 44.[1]

Hab. Java.[1]

12. SCOLIA CINCTA.

Scolia cincta, Smith, *Proc. Linn. Soc.*, II., 89, 7[1]; Sauss. and Sichel, *Cat. Scol.*, 45, 19.

Hab. Borneo, Sumatra, Sulla, Java.[1]

13. SCOLIA RUBIGINOSA.

Scolia rubiginosa, Fab. *Ent. Syst.*, II., 230, 8; *Syst. Piez.*, 241, 10; Coquebert, *Illust.*, t. 13, f. 4; Klug, Weber, u. Mohr, *Beitr.*, II., 211, 38; Lepell, *Nat. Hist. Hym.*, III., 5, 18, 2; Burm., *Cat. Scol.*, 19, 11; Smith, *Cat. Hym. Ins.*, III., 123[1]; Sauss. and Sichel, *Cat. Scol.*, 46, 20.

Scolia ornata, Lepell, *l.c.* 517, 1.

Hab. China, Siam, India, Borneo, Java, Malacca.[1]

14. SCOLIA INSIGNIS.

Scolia insignis, Sauss., *Ann. Fr. Ent. Soc.*, 1858, 197, 1; pl. v., fig. 1, ♀; Sauss. and Sichel, *Cat. Scol.*, 47, 22.[1]

Hab. Persia?[1] East Indies.[1]

16. SCOLIA DUCALIS.

Scolia ducalis, Smith, *Proc. Linn. Soc.*, V., 118, 9; Sauss. et Sichel, *Cat. Scol.*, 49.

Hab. Moluccas,[1] Ceram.[1]

15. SCOLIA CAPITATA.

Scolia capitata, Guérin, *Voy. Coq.*, 248; Burmeister, *Scol.*, 20, 13, a ♂ ; Sichel and Sauss., *Cat. Scol.*, 47.
Scolia ruficeps., Smith, *Cat.*, 111, 126.[1]
Hab. Philippines.

16. SCOLIA HÆMORRHOIDALIS.

Scolia hæmorrhoidalis, Fab., *Mant.*, I., 280, 7 ; Lep., *Nat. Hist. Ins. Hym.*, III., 552, 5 ; Burm. *Scol.*, 187 ; Smith, *Cat.*, 110, 119 ; Sauss. and Sichel, *Cat. Scol.*, 50.

This is a Palæarctic species only known from our region on the authority of Fabricius.

II. *Two cubital cellules* = Discolia

17. SCOLIA HUMERALIS.

Saussure and Sichel, *Cat.*, 321.[1]
Hab., Singapore.[1]

18. SCOLIA SCAPULATA.

Gribodo, *Ann. d. Museo Civico di Storia Nat. di Genova*, 1.
Hab. Burma.

19. SCOLIA CEPHALOTES.

Scolia cephalotes, Burmeister, *Mon. Scol.*, 37, 60;[1] Smith, *Cat. Hym.*, III., 90, 20 ; Saussure, *Stett. Ent. Zeit.*, 1859, 184?[2] Sauss. and Sichel, *Cat. Scol.*, 102, 90.
Hab. Java,[1] Borneo.[2]

20. SCOLIA CYANIPENNIS.

Scolia cyanipennis, Fab., *Syst. Piez.*, 244, 35;[1] Burm., *Mon. Scol.*, 37, 59 ; Smith, *Cat. Hym.*, III., 90, 21 ; Sauss., *Ann. Ent. Soc. Fr.*, 1858, 209, 16; Sauss. and Sich., *Cat. Scol.*, 103, 91.[2]
Hab. Java, Ceylon.[2]

21. SCOLIA COERULANS.

Scolia coerulans, Lep., *Nat. Hist. Hym. Ins.*, III., 526-7[1];
Sauss. and Sichel, *Cat. Scol.*, 104, 92.
Hab. "East Indies."[1]

22. SCOLIA MELANOSOMA.

Scolia melanosoma, Sauss., *Stett. Ent. Zeit.*, 1859, 185[1];
Sauss. and Sichel, *Cat. Scol.*, 105, 94.
Hab. Java.[1]

23. SCOLIA REDTENBACHERI.

Scolia Redtenbacheri, Sauss., *Stett. Ent. Zeit.*, 1859, 186[1];
Sauss. and Sichel, *Cat. Scol.*, 105, 95.
Hab. Java,[1] Barrackpore.

24. SCOLIA CARBONARIA.

Scolia carbonaria, Sauss., *Ann. Ent. Fr.*, 1858, 210, 17[1];
Sauss. and Sichel, *Cat. Scol.*, 106, 96.
Hab. "East Indies,"[1] Java.[1]

25. SCOLIA AUREIPENNIS.

Scolia aureipennis, Lep., *Nat. Hist. Hym.*, III., 523, 9[1];
Sauss. and Sichel, *Cat. Scol.*, 109, 102.[2]
Scolia Jurinei, Sauss., *Mélan. Hym.*, 45, 21.
Scolia instabilis, Smith, *Cat. Hym.*, III., 88, 11.
Scolia ruficornis, Klug, Weber u. Mohr, *Beitr.*, I., 25, 8.
Hab. "East Indies," Java,[2] Poona.

26. SCOLIA ERRATICA.

Scolia erratica, Smith, *Cat. Hym.* III., 88, 10;[1] *Linn. Soc.*, 88, 1 ; *l.c.*, 9, 1 ; Sauss. and Sichel, *Cat. Scol.*, 110, 103.
Scolia verticalis, Burm., *Mon. Scol.*, 37, 61 (*nec.* Fab.)[1].
„ *westermanni*, Sauss. and Sichel, *Ann. Ent. Fr.*, 1858, 212, 19.[2][3]
Hab. Java,[1] Borneo,[2] Sumatra ;[3] "East Indies."[1]

27. SCOLIA MOLESTA.

Scolia erratica, Sauss. (*nec.* Smith), *Ann. Ent. Fr.*, 1858, 211, 18; *Stett. Ent. Zeit.*, 1859, 187.

Scolia molesta, Sauss. and Sichel, *Cat. Scol.*, 111, 104.[1]

Hab. Pulvo-Penang,[1] Siam,[1] Singapore,[1] Sumatra,[1] Java,[1] Borneo.[1]

28. SCOLIA VOLLENHOVENI.

Scolia vollenhoveni, Sauss., *Stett. Ent. Zeit.*, 1859, 188; Sauss. and Sichel, *Cat. Scol.*, 112, 105.[1]

Hab. Sumatra.[1]

29. SCOLIA OBSCURA.

Scolia obscura, Lep., *Nat. Hist. Hym. Ins.*, III., 527, 14; Smith, *Cat. Hym.*, III., 89, 16.[1]

Hab. East Indies.[1]

30. SCOLIA QUADRIPUSTULATA.

Scolia 4-pustulata, Fab., *Spec. Ins.*, I., 453, 13; *Ent. Syst.* II., 234, 6; Burm., *Mon. Scol.*, 36, 58; Lep., *Nat. Hist. Hym. Ins.*, 528, 16; Smith, *Cat. Hym.*, III., 87, 7; Sauss. and Sichel, *Cat. Scol.*, 113, 108.

Larra 4-pustulata, Fab., *Syst. Piez.*, 244, 34.

Scolia binotata, Fab., *Syst. Piez.*, 244, 36.

Scolia bipunctata? Klug, Weber u. Mohr, *Beitr.* I., 36, 32.

Scolia 6-pustulata, Klug, *l.c.* 35, 30, var. ♂

Scolia fasciatopunctata, Guérin, *Voy. d. Coq.*, II., 254.

Scolia fervida, Smith, *Ann. Mag. Nat. Hist.*, IX., 46; *Cat. Hym.*, 89, 15.

Hab. Barrackpore, Bombay, Java, Sumatra.

31. SCOLIA BENGALENSIS, *sp. nov*.

Black; the flagellum of the antennæ red; two small on the second and two larger yellow marks on the third abdominal segments. Clypeus impunctate, except a row of pustules round the apex; the space above and between the antennæ very strongly and coarsely punctured; the vertex with groups of punctures round the ocelli, above these and almost impunctate to near the edge, which is closely and finely punctured. Thorax closely covered with long black hair and strongly and coarsely punctured all over, except the edges of the pleuræ. Apex of median segment transverse with a sharply oblique slope. Abdomen covered all over with widely separated punctures; the ventral segments with the punctures stronger, but with the base impunctate; the hair thick, longish and black. The ♂ is similar but not quite so strongly punctured; the third joint of the antennæ is shorter than the fourth, and the yellow marks are on the third and fourth abdominal segments.

Length, 25mm.

Hab. Poona (*Wroughton*).

Comes near to 4-*pustulata*; but that species has the antennæ black; the marks on the abdomen red, not yellow; and the thorax is not punctured all over.

In one example of *bengalensis* there is, in the ♀, two small marks on the fourth abdominal segment.

32. SCOLIA BILUNATA.

Scolia bilunata, Sauss., *Ann. Ent. Fr.*, 1858, 212, 20; Sauss. and Sichel, *Cat. Scol.*, 115, 109.

Hab. Nepaul.[1]

33. SCOLIA BIOCULATA.

Scolia bioculata, Sauss., *Stett. Ent. Zeit.*, 1859, 189; Sauss. and Sichel, *Cat. Scol.*, 115, 110.'

Hab. Java, Sumatra.[1]

34. SCOLIA FULVIFRONS.

Scolia fulvifrons, Sauss, *Mélang. Hym.*, 43, 19, f. 11; Sauss. and Sichel, *Cat. Scol.*, 116, 111.[1]

Scolia personata, Smith, *Cat. Hym.*, 91, 23.

Scolia bipunctata? Klug, Weber u. Mohr, *Beitr.*, I., 36, 32.

Hab. East Indies.[1]

[35. SCOLIA SPLENDIDA.

Scolia splendida, Sauss., *Ann. Ent. Fr.*, 1858, 213, 21, pl. V., f. 2; Sauss. and Sichel, *Cat. Scol.*, 116, 112.[1]

Hab. Asia[1] (India)?].

36. SCOLIA NOBILIS.

Scolia nobilis, Sauss., *Ann. Ent. Fr.*, 1858, 214; Sauss. and Sichel, *Cat. Scol.*, 117, 113.[1]

Hab. East Indies.[1]

37. SCOLIA SPECIFICA.

Scolia specifica, Smith, *Cat. Hym. Ins.*, III., 89, 13[1]; Sauss. and Sichel, *Cat. Scol.*, 117, 114.

Hab. East Indies.[1]

38. SCOLIA STIZUS.

Scolia stizus, Sauss. and Sichel, *Cat. Scol.*, 118, 115.[1]

Hab. Tranquebar,[1] Poona (*Wroughton*).

39. SCOLIA NITIDULA.

Scolia nitidula, Sauss., *Ann. Ent. Fr.*, 1858, 215, 23; Sauss. and Sichel, *Cat. Scol.*, 119, 117.[1]

Hab. Java.[1]

40. SCOLIA INDICA.

Scolia indica, Sauss., *Mélang. Hym.*, 46, 22, f. 10; Sauss. and Sichel, *Cat. Scol.*, 119, 118.

Scolia ignita, Smith, *Cat. Hym.* III., 101, 77.

Hab. Bengal, Silbet.

41. SCOLIA ELIFORMIS.

Scolia eliformis, Sauss. *Ann. Ent. Fr.* 1858, 215, 24[1]; Sauss. and Sichel, *Cat. Scol.*, 120, 119.

Hab. East Indies,[1] Ceylon.[1]

42. SCOLIA VENUSTA.

Scolia venusta, Smith, *Cat. Hym.*, III., 90, 17 ; Sauss. and Sichel, 120, 120.[1]

Hab. East Indies.[1]

43. SCOLIA HISTRIONICA.

Scolia histrionica, Fab., *Ent. Syst. Suppl.*, 256, 35 ; Klug, Weber u. Mohr, *Beitr.*, I., 25, 9? ; Sauss. and Sichel, *Cat. Scol.*, 121, 121.

Scolia Picteti, Sauss., *Mél. Hym.*, 42, 18.
Scolia pulchra, Smith, *Cat. Hym.*, III., 88, 12.[1]

Hab. India, Poona (*Wroughton*).

44. SCOLIA DECORATA.

Scolia decorata, Burm., *Mon. Scol.*, 30, 39 ; Sauss. and Sichel, *Cat. Scol.*, 122, 122.[1]

Scolia flavopicta, Smith, *Cat. Hym.*, III., 91, 22.

Hab. Sumatra.

45. SCOLIA VIVIDA.

Scolia vivida, Smith, *Cat. Hym.*, 89, 14 ;[1] Sauss. and Sichel, *Cat. Scol.*, 123, 125.[1]

Hab. Madras,[1] Poona (*Wroughton*).

46. SCOLIA MODESTA.

Smith, *Cat. Hym.*, III., 91, 25 ;[1] Sauss. and Sichel, *Cat.*, 124, 126.

Hab. Philippines.[1]

ELIS.

I. *With two closed cubital cellules* = Dielis.

1. ELIS AZUREA.

Elis azurea, Saussure, *Stett. Ent. Zeit.*, 1859, 269[1]; Sauss. and Sich., 185, 194.
Hab. Java, Sumatra.[1]

2. ELIS BICOLOR.

Elis bicolor, Saussure, *l.c. Ann. Soc. Ent. Fr.*, 1858, 233, 46, pl. v., f. 4;[1] Sauss. and Sich. 156, 195.
Hab. Java.[1]

3. ELIS MARGINELLA.

Elis marginella, Klug, Weber u. Mohr, II., 214, 44;[1] Sauss. and Sich., 186, 196.
? *Colpa parvula*, St. Farg., *Hym.*, III., 548, 17.
? *Scolia hirtella*, Klug, *l.c.* 215, 45.
Hab. East Indies.[1]

4. ELIS THORACICA.

Tephia thoracica, Fab., *Ent. Syst. Supp.*, 254, 15; *Syst. Piez.*, 235, 19.
Sphex albicollis, Christ, *Hymen.*, 260, t. 26, f. 1.
Sphex flavifrons, Christ, *l.c.* 261, t. 26, f. 2.
Tiphia nigra, Fab., *Ent. Syst.*, II., 225, 9; *Syst. Piez.*, 234, 13.
Campsomeris auricollis, Lep., *Hym.*, III., 499.[6]
Elis thoracica, Saus. and Sich., *Cat.*, 188, 197.
Hab. Java, China, Barrackpore, Poona.

5. ELIS FIMBRIATA.

Elis fimbriata, Burmeister, *Scol.*, 25, 6; Sauss. and Sich., 189, 198.
Scolia thoracica, Klug, Weber, and Mohr, I., 33, 24.
Campsomeris collaris, Lepel., *Hym.*, III., 498, 5.
Hab. Java.

6. ELIS ASIATICA.

Elis Asiatica, Saussure, *Ann. Soc. Ent. Fr.*, 1858, 231, 34;[1] *Ent. Zeit*, 1859, 266; Sauss. and Sichel, *Cat.*, 190, 200.
Hab. Java, East Indies.[1]

7. ELIS RETICULATA, sp. nov.

Black, the wings fusco-violaceous. Clypeus coarsely punctured at the base; the space above and between the antennæ coarsely and closely punctured; the vertex with a few scattered punctures, behind the ocelli with the punctures closer together and more numerous. The entire head thickly covered with long black hairs. Thorax closely and strongly punctured all over, except the apex of the scutellum and the apex of the mesopleuræ, and the base of the metapleuræ: the pronotum transverse in front; the apex of the median segment, almost transverse; without furrows and with an oblique slope, and punctured closely all over; the entire thorax bearing long black hair. Abdomen shining, having a bluish tinge, sparsely punctured and densely black haired all over; the apex of the anal segment impunctate; the ventral segments sparsely punctured all over. Antennæ dull black; the third and fourth joints subequal; the apical joints dilated beneath. ♂.

Length 19 mm.

Hab. Poona (*Wroughton*).

Comes near to *Javana* but that species has in the ♂ the abdominal segments cinereo-ciliated, and the mesonotum impunctate in the middle.

8. ELIS JAVANA.

Elis Javana, Lepel., *Hym.*, III., 498, 402;[1] Sauss. and Sich., *Cat.*, 191, 202.

Hab. Java.[1]

9. ELIS TRISTIS.

Elis tristis, Saussure, *Ent. Zeit.*, 1859, 265;[1] Sauss. and Sich., *Cat.*, 193, 205.

Hab. Java, Borneo, East Indies.[1]

10. ELIS LUCTUOSA.

Elis luctuosa, Smith, *Cat. Hym.*, 101, 77 (*Scolia*); Sauss. and Sich., *Cat.*, 194, 206.

Scolia 4-guttulata, Sauss., *Mél. Hymén.*, 58, f. 12.

Hab. India, Java, Philippines.

11. ELIS QUADRIGUTTULATA.

E. quadriguttulata, Burmeister, *Scol.*, 21. 17 (*Scolia*);[1] Sauss. and Sich., *Cat.*, 195, 207.

Hab. Java.[1]

12. ELIS EXIMIA

Elis eximia, Smith, *Cat. Hym.*, III., 99, 69 (*Scolia*);[1] Sauss and Sich., *Cat.*, 195, 208.

Hab. India.[1]

13. ELIS RUBROMACULATA.

Elis rubromaculata, Smith, *Cat. Hym.*, III., 99, 67 (*Scolia*); Sauss. and Sich., *Cat.*, 196, 209.

Hab. Java.

14. ELIS ANNULATA.

Elis annulata, Fab., *Ent. Syst.*, II., 225, 7 (*Tiphia*); Burmeister, *Scol.*, 25, 27; Sauss. and Sichel, *Cat.*, 196, 210.

Campsomeris Servillii, Lepel, *Hym.*, III., 501, 9.

Hab. China, Japan, Barrackpore, Poona, Burma, Java, Manilla.

15. ELIS DREWSENI.

E. Drewseni, Sauss., *Ann. Ent. Fr.*, 1858, 232, 44;[1] Sauss. and Sich., *Cat.*, 197, 211.

Hab. Java.[1]

16. ELIS HABROCOMA.

E. habrocoma, Smith, *Cat. Hym.*, III., 100, 71[1] (*Scolia*); Sauss. and Sichel, *Cat.* 198, 212.

Hab. India.[1]

17. ELIS SNELLENI.

E. Snelleni, Sauss., *Stett. Ent. Zeit.*, 1859; 268, tab. 2, f. 4;[1] Sauss. and Sichel, *Cat.*, 198, 213.

Hab. Sumatra.[1]

18. ELIS GROSSA.

E. grossa, Fab., *Syst. Piez.*, 232, 4 (*Tiphia*); Burmeister, *Scol.*, 23, 22 (*Scolia*); Sauss. and Sich., *Cat.*, 199, 215.

Elis sericea, Sauss., *Mél. Hym.*, 63, 31.

Hab. India, Java.

19. ELIS HIRSUTA.

Elis hirsuta, Sauss., *Ann. Ent. Fr.*, 1858, 234, 47; Sauss. and Sich., *Cat.*, 200, 216.

Hab. Tranquebar.

20. ELIS IRIS.

Elis Iris, Lepel, *Hym.*, III., 547, 16; Sauss. and Sichel, *Cat.*, 201, 217.

Elis phalerata, Sauss., *Ann. Ent. Fr.*, 1858, 233, 45.

Hab. Java.

21. ELIS CEYLONICA.

Campsomeris ceylonica, Kirby, *Trans. Ent. Soc.*, 1889, 452.

Hab. Ceylon.[1]

22. ELIS HINDENII.

Elis hindenii, Lepel, *Hym.*, III., 500. 8 (*Campsomeris*); Sauss. and Sich , *Cat.*, 204, 219.

Scolia quadrifasciata, Fab., *Ent. Syst. Supp.*, 255, 16-17.

Hab. Japan, China, India, Moluccas.

23. ELIS LIMBATA.

Elis limbata, Sauss. *Cat.*, 206, 220.[1]

Hab. Java.[1]

24. ELIS AURULENTA.

Elis aurulenta, Smith, *Cat. Hym.*, III., 206 ; Sauss. and Sich., *Cat.*, 206, 221.

Hab. Philippines, Celebes, Bachian.

25. ELIS CYANEA.

Saussure, *Cat.*, 323.

Hab. Nicobar Islands.

26. ELIS LITIGIOSA.

Elis litigiosa, Smith, *Cat. Hym.*, III., 113, 133;[1] Sauss. and Sich., 158, 164.

Hab. East Indies.[1]

II. *With three cubital cellules* = Triclis.

27. ELIS ORIENTALIS, *sp. nov.*

Black; the wings dark violaceous. The clypeus impunctate, with a row of punctures round the apex; the vertex impunctate, except a semi-circle of punctures round the hinder ocelli to near the eye incision; front strongly punctured above the antennæ, and there is a row of punctures at the edge of the vertex, the front densely covered with blackish, the cheeks and outer orbits with greyish, hair. Thorax moderately strongly punctured, except on the mesopleuræ behind the margin of the propleuræ, the metapleuræ at the base, and the centre of the mesonotum; the latter having the punctures sparser; and with the parapsidal furrows deep, and reaching from the apex to beyond the middle. Scutellum sparsely punctured; the sides of the metanotum very closely rugosely punctured. Median segment short, transverse at the apex, which has a rather sharply oblique slope and is impunctate between the furrows. Abdomen shining, sparsely punctured, the hair black, longish, and sparse; pygidial area coarsely longitudinally rugose; the ventral segment sparsely punctured on the apical half and bearing long black hairs. The third cubital cellule of nearly equal width throughout. The hair on the legs is black, except on the fore femora, which have it greyish; the hind femora have behind a row of punctures in the middle.

Length 17 mm.

Hab. Ceylon (*Rothney*).

MYZINE, *Latr.*

In his *Catalogue of Indian Hymenoptera*, Smith omits, curiously enough, all the species described by himself in his *Cat. Hym.*, III.

1. MYZINE ANTHRACINA.

Smith, *Cat. Hym.*, III., 71, 9.

Hab. India.

2. MIZINE COMBUSTA.

Smith, *New species of Hym.*, 179.

Hab. India? or Africa?

3. MYZINE DIMIDIATA.

Guérin, *Dict. pitt. d. Hist. Nat.*, v. 575, 17; Smith, *Cat. Hym.*, 71.

Hab. Bombay, Bengal.

4. MYZINE FUSCIPENNIS.

Smith, *Cat. Hym.*, III., 72.
Hab. India.

5. MYZINE MADRASPATANA.

Smith, *Cat. Hym.*, III., 72.
Hab. Madras.

6. MYZINE NITIDA. (Pl. IV., f. 2.)

Black, shining, the abdomen with a bluish gloss; the wings fuscous, hyaline to the transverse basal nervure. Head covered with longish greyish hair, especially long and thick below the antennæ; coarsely rugosely punctured below the ocelli, the vertex with the punctures small, shallower and more widely separated; the ocellar region slightly raised, a depression on the outerside of the lateral; a not very distinct channel runs down from the anterior. Mandibles testaceous black at the apex; antennæ stout, opaque, as long as the abdomen, the first and second joints subequal, the third, if anything, longer than the fourth. The pronotum punctured, somewhat like the vertex; the mesonotum in front finely and closely punctured; behind with the punctures larger and widely separated; the scutellum slightly raised, very coarsely rugose; the median segment rugosely punctured; the mesopleura convex, coarsely punctured; behind there is a smooth, shining oblique depression. The thorax bears a thick greyish hair. Abdomen very smooth and shining; the pygidium margined laterally, carinate down the centre; the space between with some large punctures; the apical segments bearing longish blackish hairs. Legs bearing a greyish pubescence; the fore tibiæ yellowish in front. The second cubital cellule at top and bottom shorter than the third; the first recurrent

nervure received beyond, the second in front of middle of cellule. Length nearly 15 mm.
Hab. Poona (*Wroughton*).

Allied to *M. fuscipennis;* but that species differs in having "a central channel, which is margined by too slightly elevated carinæ," the abdomen strongly punctured at the base, &c.

7. MYZINE ORIENTALIS.

Smith, *New species of Hym.*, 179.
Hab. Beloochustan.

8. MYZINE PALLIDA.

Smith, *New Species of Hym.*, 179.
Hab. North West Provinces.

9. MYZINE PETIOLATA.

Smith, *Cat. Hym.*, III., 72.
Hab. Poona (*Wroughton*).

10. MYZINE TRICOLOR.

Smith, *Proc. Linn. Soc.*, II., 91, 1.
Hab. Borneo.[1]

TIPHIA, *Fab.*

1. TIPHIA COMPRESSA, Smith, *Cat. Hym.*, III., 82, 4.
Hab. Philippines.

2. TIPHIA HIRSUTA, Smith, *l.c.* 83, 5.
Hab. North India.

3. TIPHIA RUFIPES, Smith. *l.c.* 83, 6.
Hab. North India.

4. TIPHIA RUFO-FEMORATA, Smith, *l.c.* 83, 7.
Hab. North India.

5. Tiphia fumipennis, Smith, *Proc. Linn. Soc.*, II., 90. 1.
Hab. Borneo.

6. Tiphia consueta, Smith, *New Species of Hymenoptera*, 184.
Hab. Ceylon.

MUTILLIDÆ.

MUTILLA, *Linn.*

1. Mutilla accedens, Radoszkovsky and Sichel, *Mon. d. Mutill*, 89.
Hab. Manilla, Luzon.[1]

2. Mutilla aestuans, Gerstacker, *Peter's Reise*, 487, pl. 31, f. 6[1]; Radoszkovsky and Sichel, *Mongr. J. Mutill*, 85.[2]
Hab. Ceylon,[2] Mozambique,[1] Caffreria.[2]

3. Mutilla analis, Lep., *Hym.*, III., 630, 52[1]; Rad. and Sichel, *Mon. d. Mutill*, 146.[2]
Mutilla fuscipennis, Fab., *Syst. Piez.*, 436, 35.
Mutilla rufogastra, Smith, *Cat. Hym.*, 36, 185.
Hab. India,[1] Ceylon.[2]

4. Mutilla anonyma, Kohl, *Verb. z.-b. Ges. Wien.*, 1882, 482, f. 20.
Hab. Sumatra.[1]

5. Mutilla antennata, Smith, *Cat. Hym.*, III., 31, 166.[1]
Hab. India.[1]

6. Mutilla argenteomaculata, Smith, *New Species of Hym.*, 199.[1]
Hab. Bombay Presidency.[1]

7. MUTILLA ARGENTIPES, Smith, *Cat. Hym.*, III., 31, 167.[1] (Pl. IV. f. 3 ♂, f. 14 ♀).
   Hab. India,[1] Poona (*Wroughton*).

8. MUTILLA AULICA, Smith, *Cat. Hym.*, III., 37, 189[1]; Rad. and Sich., *Mon. d. Mutill.*, 120. (Pl. IV. f. 4).
   Hab. North India,[1] Poona (*Wroughton*).

9. MUTILLA AUREORUBRA, Rad. and Sich., *Mon. d. Mutill.*, 166.

   MUTILLA EGREGIA, Saussure, *Non* Klug. *Ann. Soc. Ent. Fr.*, 1867, 351. Pl. VIII. f. 1.
   Hab. Trincomalia, Ceylon.

10. MUTILLA AURIFEX, Smith, *New Species of Hym.*, 198.
    Hab. Bombay Presidency.

11. MUTILLA AURIFRONS, Smith, *Cat. Hym.*, III., 31, 168.[1]
    Hab. India.[1]

12. MUTILLA BASILIS, Smith, new species of Hym., 200.[1]
    Hab. Bombay Presidency.[1]

13. MUTILLA BENGALENSIS, Lep., *Hym.*, III., 637, 63;[1] Rad. and Sich, *Mon. d. Mutill.*, 122.
    Hab. Bengal.[1]

14. MUTILLA BICINCTA, Saussure, *Ann. Soc. Ent. Fr.* 1867[1] 355, t. 8, f. 4.[1]
    Hab. Paradinie, Ceylon.[1]

15. MUTILLA BLANDA, Smith, *Cat. Hym.*, III., 32, 170.
    Hab. India.[1]

16. MUTILLA BUDDHA, Cam. (Pl. IV., f. 9).
    Hab. Poona (*Wroughton*).

17. MUTILLA CALLIOPE, Smith, *Proc. Linn. Soc.*, II., 85, 8.[1]
    Hab. Borneo.[1]

18. MUTILLA CASSIOPE, Smith, *Proc. Linn. Soc.*, II., 86, 12.[1]
    Hab. Borneo.[1]
    109.[1]

19. MUTILLA CEYLANENSIS, Rad. and Sich., *Mon. d. Mutill*, *Hab.* Ceylon.[1]

20. MUTILLA CHRYSOPHTHALMA, Klug, *Hym. b. phys.*, 17, pl. V., f. 3., Rad. and Sich., *Mongr. Mutill*, 95.[1] *Hab.* Arabia, Ceylon.[1]

21. MUTILLA COMOTTII, Gribodo. *Hab.* Burma.

22. MUTILLA CONSTANCEÆ, Cam. (Pl. IV., f. 10) *Hab.* Poona (*Wroughton*).

23. MUTILLA CORONATA, Saussure, *Novara Reise*, 106. *Hab.* Ceylon.[1]

24. MUTILLA COROMANDELICA, Motsch., *Bull. Mosc*, 1863. 23. *Hab.* India, Madurà.

25. MUTILLA DARDANUS, Smith, *Proc. Linn. Soc.*, II., 86, 13.[1] *Hab.* Borneo.[1]

26. MUTILLA DECORA, Smith, New Species of Hym, 200.[1] *Hab.* Pulo Penang.[1]

27. MUTILLA DEIDAMIA, Smith, *Proc. Linn. Soc.*, II.,83, 3.[1] *Hab.* Borneo.

28. MUTILLA DENTICOLLIS, Motsch, *Bull. Mosc.*, 1863, 22. *Hab.* Ceylon, Mountains of Nura Ellia.

29. MUTILLA DIMIDIATA, Lep., *Hym.*, III., 628, 50;[1] Rad. and Sich., *Mon.*, 147.
Mutilla rufogastra, Lep., *l.c.*, 629, 51.
Mutilla sexmaculata, Smith, *Cat, Hym.*, III., 37, 188.
Hab. India,[1] Pondechery, Triconomale, Timor, Luzon.

30. MUTILLA DIVERSA, Smith, *Cat. Hym.*, 32, 171.[1] *Hab.* India.[1]

31. MUTILLA DIVES, Smith, *Cat. Hym.*, III., 32, 172.[1] *Hab.* India.[1]

32. MUTILLA ERYTHROCERA, Cam. *Hab.* Poona (*Wroughton*).

33. MUTILLA FAMILIARIS, Smith, *Proc. Linn. Soc.*, II., 84, 7.[1]
    Hab. Singapore, Borneo.[1]

34. MUTILLA FUNERARIA, Smith, *Cat. Hym.*, III., 37, 190.[1]
    Hab. North India.

35. MUTILLA GLABRATA, Fab., *Syst. Piez.*, 438, 45 ;[1] Olivier,
    Ent. Méth., VIII., 65, 64.
    Hab. India.[1]

36. MUTILLA GRACILLIMA, Smith, *Proc. Linn. Soc.*, II., 84, 6.[1]
    Hab. Borneo.[1]

37. MUTILLA HEXAOPS, Saussure, *Ann. Soc. Ent. Fr.*, 1867,
    169, 7.[1]
    Hab. Nattan, Ceylon.[1]

38. MUTILLA HUMBERTIANA, Saussure, *Ann. Soc. Ent. Fr.*,
    1867, 353, t. 8, f. 2.
    Hab. Trincomalia.[1]

39. MUTILLA HYMALAYENSIS, Radosykowsky, *Horae. Soc.
    Ent. Ross.*, XIX.
    Hab. Himalaya.[1]

40. MUTILLA INDICA, Linné, *Syst. Nat.*, I., 966, 3.[1]
    Hab. India.[1]

41. MUTILLA INDOSTANA, Smith, *Cat. Hym.*, III., 33, 175.[1]
    Hab. Madras.[1]

42. MUTILLA INSULARIS, Cam.
    Hab. Sober Island, Trincomalia, Ceylon (*Yerburgh*).

43. MUTILLA INTERMEDIA, Saussure, *Ann. Soc. Ent. Fr.*,
    1867, 354, 4.[1]
    Hab. Ceylon.[1]

44. MUTILLA KAUARÆ, Cam. (Pl. IV., f. 2.)
    Hab. Ceylon (*Yerburgh*).

45. MUTILLA KAUTHELLÆ, Cam.
    Hab. Kauthella, Ceylon (*Yerburgh*).

46. MUTILLA MACULOFASCIATA, Saussure, *Novara Reise, Hym.*, 107, 5.
Hab. Ceylon, Timor, Luzon.[1]

47. MAHAGANAYENSIS, Cam.
Hab. Mahaganay, Ceylon.

48. MUTILLA METALLICA, Cam.
Hab. Trincomali, Ceylon (*Yerburgh*).

49. MUTILLA MIRANDA, Smith, *Cat. Hym.*, III., 33, 176.[1]
Hab. India.[1]

50. MUTILLA NEREIS, Kohl, *Verb. z. b. ges. Wien*, 1882, 476, f. 2.[1]
Hab. Java,[1]

51. MUTILLA NIGRIPES, Fab., *Syst. Piez.*, 439, 51.
Hab. India.[1]

52. MUTILLA NOBILIS, Smith, *Cat. Hym.*, III. 33, 178.
Hab. Madras.

53. MUTILLA OCCELLATA, Saussure, *Ann. Soc. Ent. Fr.*, 1867, 169, 6.[1]
Hab Ceylon.[1]

54. MUTILLA OPTIMA, Smith, *Cat. Hym.*, III., 34, 179.[1]
Hab. India.[1]

55. MUTILLA OPULENTA, Smith, *Cat. Hym.*, III., 34, 180.[1]
Hab. India,[1] Kauthalla, Ceylon (*Yerburgh*).

56. MUTILLA PANDORA, Smith, *Proc. Linn. Soc.*, II., 85, 10
Hab. Borneo.[1]

57. MUTILLA PHILIPPINENSIS, Smith, *Cat. Hym.*, III., 40, 200;[1] Rad. and Sich., *Mongr. Mutil.*, 88.
Hab. Philippines,[1] Luzon.

58. MUTILLA PLACIDA, Smith, *New Species of Hym.*, 198.[1]
Hab. Bombay Presidency.[1]

59. MUTILLA PONDICHERENSIS, Radosykovsky and Sichel, *Monog. d. Mutilles*, 66.
   Hab. Pondichery.[1]

60. MUTILLA POONAENSIS, Cam
   Hab. Poona (*Wroughton*).

61. MUTILLA PROSPERINA, Smith, *Proc. Linn. Soc.*, II., 85, 9.[1]
   Hab. Borneo.[1]

62. MUTILLA PULCHRICEPS, *Cam.* (Pl. IV., f. 17)
   Hab. Poona (*Wroughton*).

63. MUTILLA PULCHRINA, Smith, *Cat. Hym.*, III., 34, 181,[1]
   (Pl. IV., f. 6-8 and 16).
   Hab. Madras,[1] Savoy, Poona (*Wroughton*).

64. MUTILLA PULCHRIVENTRIS, *Cam.* (Pl. IV., f. 5)
   Hab. Poona (*Wroughton*).

65. MUTILLA PUSILLA, Smith, *Cat. Hym.*, III., 37, 191.[1]
   Hab. North India.[1]

66. MUTILLA REGIA, Smith, *Cat. Hym.*, III., 38, 192[1] (Pl. IV., f. 7).
   Hab. North India,[1] Poona (*Wroughton*).

67. MUTILLA RUFITARSIS, Smith, *New Species of Hym.*, 199.[1]
   Hab. India.[1]

68. MUTILLA RUFIVENTRIS, Smith, *Cat. Hym.*, III., 36, 184.[1]
   Hab. India.[1]

69. MUTILLA RETICULATA, *Cat. Hym.*, III., 35, 183.[1]
   Hab. India.[1]

70. MUTILLA RUGOSA, Olivier, *Ency. Méth.*, VIII., 61, 35; Rad. and Sich., *Mon. d. Mutill.*, 121. (Pl. IX. f. 4).
   Hab. India,[1]

71. MUTILLA SEMIAURATA, Smith, *Cat. Hym.*, III., 36, 187.
   Hab. India.[1]

72. MUTILLA SERRATULA, Cam. (Pl. IV., f. 12).
   Hab. Poona (Wroughton).

73. MUTILLA SEXMACULATA, Swed., Nov. Act. Holm., VIII., 286, 44; Rad. and Sich., Mon. d. Mutill, 109[1] (non Smith).
   Hab. Bengal,[1] South India, Calcutta.

74. MUTILLA SIBYLLA, Smith, Proc. Linn. Soc., II., 86, 11.
   Hab. Borneo, Celebes, Arn.[1]

75. MUTILLA SUBINTRANS, Rad. and Sich., Mon. d. Mutill, 90.[1]
   Hab. Ceylon, Timor.[1]

76. MUTILLA SOROR, Saussure, Ann. Soc. Ent. Fr., 1867, 354, t. 8, f. 3.[1]
   Hab. Habourenne, Ceylon.[1]

77. MUTILLA SUSPICIOSA, Smith, Proc. Linn. Soc., II., 94, 5.[1]
   Hab. Borneo, Celebes, Amboyna, Bourn, Flores.[1]

78. MUTILLA TETRAOPS, Rad. and Sich., Mon d. Mutill. 119.[1]
   Mutilla leucopyga, Smith, Cat Hym., III., 12, 74 (nec Klug).
   Mutilla sexmaculata, Smith, l.c. 37, 188 (nec Swed.).[2]
   Hab. China, India.[2]

79. MUTILLA TRIMACULATA, Cam.
   Hab. Poona (Wroughton).

80. MUTILLA TRINCOMOMALICA, Rad.
   Hab. Trincomalia.[1]

81. TROPBAN.E, Cam.
   Hab. Trincomalia (Yerburgh).

82. MUTILLA UNIFASCIATA, Smith, Cat. Hym., III., 38, 193.
   Hab. India, Celebes.

   ? MUTILLA VICINISSIMA, Gribodo, Ann. d. Mus. Civ. a. Storia Nat. d. Genova.
   Hab. Burma.

83. MUTILLA UNIMACULATA, Smith, *Proc. Linn. Soc.*, II., 87, 14.
Hab. Borneo, Celebes.[1]

84. MUTILLA URANIA, Smith, *Proc. Linn. Soc.*, II., 83, 4.[1]
Hab. Borneo.[1]

85. MUTILLA VEDA, Cam.
Hab. Poona (*Wroughton*).

86. MUTILLA WROUGHTONI, Cam. (Pl. IV. f. 15).
Hab. Poona (*Wroughton*).

87. MUTILLA YERBURGI, Cam.
Hab. Mahaagang, Ceylon (*Yerburgh*).

The following table may enable the new species here described to be more easily recognised :—

1. (2) Abdomen without spots. The apex of the second and the third abdominal segments with golden bands; the head covered densely with a golden pubescence; thorax elongate-quadrate, concave, dilated at the apex. *Kafara.*
2. (1) Abdomen with spots.
3. (14) The spots golden.
4. (9) Head black.
5. (6) Sides of thorax irregularly serrulate; tibiae and tarsi testaceous. Second abdominal segment without a large golden spot. Length 5 mm. *serratula.*
6. (5) Sides of thorax not irregularly serrulate; tibiae and tarsi black.
7. (8) Thorax laterally concave, dilated at apex, abdomen at base with two large dilated marks, legs not densely pilose. *insularis.*
8. (7) Thorax laterally not concave, narrowed towards the apex, abdomen at base with a small spot, legs densely pilose. *Buddha.*
9. (4) Head red.
10. (13) Abdomen with two golden spots; the sides of thorax not irregularly serrulate.

11. (12) Head large, more than half the length of the thorax; the pubescence sparse, fuscous. *erythrocera.*
12. (11) Head small, less than half the length of the thorax, the pubescence dense, golden. *pulchriceps.*
13. (10) Abdomen with one small golden spot at the base; the sides of thorax irregularly serrate. *veda.*
14. (3) The spots white.
15. (20) The head red, wholly or in part.
16. (17) Abdomen with one spot and one band. *Poonaensis.*
17. (16) With two or three spots and no band; abdomen purple.
18. (19) With two spots, thorax laterally slightly convex, not much narrowed towards the apex. *metallica.*
19. (18) With three spots, thorax laterally concave in middle, distinctly narrowed towards the apex. *pulchriventris.*
20. (15) Head black.
21. (24) Abdomen with no white band.
22. (23) ,, ,, two spots placed transversely. *Wroughtoni.*
23. (24) ,, ,, three spots placed longitudinally. *trimaculata.*
24. (21) ,, ,, a band or two.
25. (26) ,, ,, one spot, the thorax dilated at apex. *Taphanae.*
26. (25) ,, ,, two or three spots, the thorax not dilated at apex.
27. (28) With two spots; thorax without spines. *Constanceae.*
28. (27) With three spots; thorax more than twice the length of the head, spined. *Kauthellae.*

MUTILLA KAUTHELLAE, sp. nov.

Black, the mesonotum dull red; a broad macula on the apex of the first segment, two oval ones on the second; and two irregular ones on the apex of the third, white; the ventral segments fringed with white hairs. Head large, distinctly wider than the thorax, very coarsely, irregularly rugosely punctured, forming almost reticulations beneath. Antennal tubercles obliquely striated. The hair on the vertex is fulvous, intermixed with longer fuscous hair;

on the lower part of the head pale to silvery. Eyes oval, convex; vertex convex, raised above the top of the eyes. Scape of the antennæ punctured, covered with glistening, silvery hairs; the flagellum thick, tapering towards the apex, where it is brownish; the third joint more than twice the length of the fourth, and longer than the fourth and fifth united. Thorax elongate quadrate, with the punctures larger and deeper than on the head, the mesopleuræ almost opaque, aciculate; above the line of separation, between the meso—meta—and median, segment is obliterated; the prothorax is very short; the mesothorax becomes a little dilated towards the apex; the sides at the apex with two obtuse tubercles; the base being also dilated, so that the central region is narrower than the basal and apical. On the base of the median segment, at the side, is a large, shining, oblique tooth. Abdomen deep black, at base subsessile; longer than the head and thorax united; above thickly covered with black hairs; there is a line of dark, fulvous hairs on the side of the second segment; pygidium finely rugose, covered with long brownish hairs. Legs covered with long white hairs; the calcaria pale yellowish-testaceous; the tibiæ with stout spines.

Length, 12mm.

MUTILLA TAPROBANE, *sp. nov.*

Black, the thorax ferruginous; an ovate mark on the second segment in the middle, a square one at its apex, and the third segment silvery-white. Head as wide as the thorax, coarsely rugosely punctured; sparsely covered with long pale hairs; convex above, but not much developed above the eyes; antennal tubercles smooth shining. Scape of antennæ and the second joint piceous; the third joint more than twice the length of the fourth, and as long as the fourth and fifth united. Thorax punctured like the head, quadrate; the hairs long and black; the sides very slightly concave, rough, hardly dilated towards the apex. Mesopleuræ shining, impunctate in the middle, the metapleuræ the same except at the apex. Median segment rounded, and with a very slight slope. Abdomen as long as the head and thorax united; subpetiolate; less strongly punctured than the thorax; the hairs at the base black, at apex pale; the ventral segments fringed with long silvery hair; pygidium longitudinally striated, rufous in the centre. Legs bearing whitish hairs; the calcaria pale; the tibial spurs thick and of four pairs. Mandibles piceous at the base.

Length, 7 mm.

M. rufitarsis, Sm., appears to have pretty much the same colouration and markings as the above, but it is larger (4½ lines).

MUTILLA TRIMACULATA, *sp. nov*.

Black, the thorax dark rufo-testaceous, a round mark on the base and apex of the second segment, and a larger longish one on the fourth, glistening white. Antennæ stout, the flagellum dull brownish beneath, the scape obscure rufo-testaceous, the third joint clearly longer than the fourth. Head as wide as the thorax, rather strongly punctured; covered with long blackish hairs; antennal tubercles shining, dull rufo-testaceous, this being also the colour of the mandibles at the apex. Thorax somewhat more strongly punctured than the head; the pleuræ impunctate; the sides of the thorax above straight, becoming slightly, but distinctly, narrowed from base to apex; the median segment with a rather sharply rounded slope. Abdomen not much longer than the head and thorax united, the pygidium shining, finely punctured. The ventral segments and pleuræ dull piceous, the thorax and abdomen bearing long blackish hairs; the apical ventral segments fringed with whitish hairs; the legs bear white hairs; the fore tarsi piceous; calcaria pale. Nearly related to *M. metallica*, but wants the metallic gloss on the thorax and head; the abdomen black, the base of the second segment with a white mark, the fascia on its apex being continuous; the third antennal joint shorter in proportion to the fourth; the thorax more distinctly narrowed from base to apex; the median segment with the slope more gradual, not so sharply oblique.

Length, nearly 5 mm.

MUTILLA WROUGHTONI, *sp. nov.*

Black, the thorax above rufous, the base of the second abdominal segment with two oval white marks, Antennæ stout; the third joint about one-half longer than the fourth. Head broader than the thorax, coarsely rugosely punctured; eyes moderate, oblong, the head well developed behind them. Thorax more coarsely rugose than the head, the pleuræ apparently impunctate; the sides of the thorax above rough, becoming gradually dilated to the apex; the apex of median segment oblique, black. Abdomen shorter than the head and thorax united; the subsessile pygidum apparently punctured, covered with long hairs. The upper surface of the insect has the hair black; the ventral longer and whitish. Legs covered with white hairs.

Length, 8½ mm.

MUTILLA PULCHRIVENTRIS, *sp. nov.*

Head and antennæ red, the latter covered on the top thickly with pale golden pubescence, hiding the ground colour; thorax dull red; abdomen a small spot longer than broad at the base and two broader than long on the apex of the second segment, white; legs red; the femora and tibiæ more or less purple. Head wider than the thorax; eyes large, oval, reaching quite close to the top of the head. The third antennæ joint as long as the fourth and fifth united. Thorax elongate, rounded at base and apex concave near the middle, distinctly narrowed towards the apex, above coarsely punctured; the median segment with the punctures larger, rounder, deeper, and more widely separated; pleuræ coppery, the meso covered thickly with white hair; the mesonotum with long black hair, and with a short white glistening sparse pubescence. Abdomen oval, wider than the thorax, narrowed at base and apex, closely punctured and bearing long black hairs; pygidium impunctate; from the apex of the penultimate segment spring two masses of white hair; ventral segments filled with long pale hair. Legs covered with pale hair; there are three rows of tibial spines; the calcaria white.

Length, 9 mm.

Very closely related to *metallica*, but easily separated by the different shape of the abdomen, which in *M. metallica*, is longer, narrower, not dilated in the centre, but becomes gradually narrower towards the apex; it has also the punctuation on the median segment as on the mesonotum.

Mutilla metallica, *sp. nov.*

Head and thorax coppery, metallic, with greenish tints; abdomen bright metallic purple; the legs, antennæ, and oral region widely pale ferruginous. Head slightly wider than the antennæ; shining, moderately strongly punctured; the oral region with a glistening white pubescence; the upper part with long blackish hairs. The second and fourth joints of the antennæ about subequal; half the length of the third. Thorax above, somewhat more strongly punctured than the head; the propleuræ in the middle punctured; the mesopleuræ in the middle aciculate; above the middle legs there is a small space finely longitudinally striated. The sides of the mesothorax almost straight, but very slightly narrowed towards the apex; the median segment with a gradual rounded slope to the apex, where it is distinctly narrowed. Abdomen as long as the head and thorax united, more closely punctured than the thorax; at the base, narrowed, narrower than the apex of the median segment; it bears longish black hairs; and they become longer in the apical segments; the ventral segments fringed with pale hairs; on the apex of the second segment are two spots of white hairs, broader than long; pygidium finely punctured. Legs covered with long white hairs; the tibiæ and femora more or less metallic coppery green behind. Eyes large, oval, convex; head not much convex on top. There are three tibial spines.

The quantity of ferruginous on the head varies; the extreme base of the thorax and abdomen are also of this colour.

Length, 7 mm.

— MUTILLA POONAENSIS, *sp. nov*.

Head, thorax, base of abdomen rufo-testaceous, abdomen black, with a purplish gloss, a small white mark on the second segment near the base and a band on its apex, glistening white. Antennæ of moderate length, stout, the third joint not much longer than the fourth. Head a little broader than the thorax, closely and coarsely punctured, covered with white glistening hairs; the antennal tubercles impunctate; eyes large, oblong, reaching quite close to the head. Thorax much more coarsely punctured than the head; the prothorax rounded in front; the median segment with a gradual rounded slope to the apex, coarsely punctured; mesopleuræ shining, impunctate; the thorax covered with longish, fuscous to glistening white hairs. Abdomen subpetiolate, narrowed at the base; pygidium impunctate, covered with long fuscous and white hairs. Legs covered with long white hairs; the calcaria white. The sides of the thorax slightly narrowed from base to apex.

Length, 6 mm.

MUTILLA VEDA, sp. nov.

Head and thorax ferruginous; abdomen black; the extreme base ferruginous; a mark, broader than long, on the base, and a large band (dilated in the centre) on the apex of the second segment, pale fulvo-golden; the scape and the basal joints of the flagellum and the legs reddish; the apical joints of the flagellum brownish beneath; the apices of the femora broadly blackish.

Very closely allied to *M. serratula*, having the same form of the thorax; differing in having the head red, and in there being a spot on the base of the second segment, in the pygidium being covered all over with pale fulvous hairs; in the thorax at the base laterally being not so rounded; the sides at the extreme base project into a broad tooth, roundly incised at the apex; the sides in the middle are not contracted, nor continuously serrate, but broadly waved; the apex of the median segment is rounded, not brought to a point in the middle and serrate, the metapleuræ are distinctly punctured. Legs covered with longish hair. The third joint of the antennæ a little longer than the fourth. Eyes moderate, oblong; clypeus covered with long fulvous hairs; the base of the mandibles rufo-piceous.

Length, 5 mm.

MUTILLA PULCHRICEPS, sp. nov.

Thorax ferruginous, the scape and base of the flagellum and legs rufo-testaceous: abdomen black, two large round maculæ on the second, and the third and fourth segments golden-fulvous; the head apparently black, but hid by a thick covering of fulvous pubescence. Head coarsely punctured; convex above, not much developed above the top of the eyes. Antennal tubercles coarsely punctured, eyes large, oblong. Second antennal joint shorter than the fourth, the third about twice the length of the fourth.

Thorax coarsely punctured; the hairs long and black; the pleuræ shining impunctate. Sides of thorax from above distinctly concave, irregular; the apex hardly wider than the base; the sides of median segment serrate. Median segment with an abruptly oblique slope. Abdomen as long as the head and thorax united, subpetiolate. Pygidium longtudinally rugose and covered with long pale fulvous hairs. The ventral segment fringed with golden fulvous hairs. The coxæ and apex of the femora are black; the hairs on femora are pale; on tibiæ and tarsi pale rufous; the tibial spines five in number, longish, and like the calcaria, pale rufous. Basal half of mandibles and antennal tubercles rufo-piceous.

Length, 8 mm.

Allied to *M. soror* and *intermedia*, with which it agrees in colouration; but differs in the head, being densely covered with a golden fulvous pubescence. *Soror* is further distinguished from it in the thorax, not being concave, but straight; and *intermedia* has the metapleuræ rugose.

—MUTILLA ERYTHROCERA, *sp. nov.*

Antennæ, head, thorax and legs, for the greater part, ferruginous; abdomen black, two large round maculæ on the second segment, and the third and fourth segments golden-fulvous. Head coarsely rugosely punctured, and sparsely covered with long blackish hairs; the hairs on the clypeus pale fulvous; antennal tubercles shining, impunctate. Palpi testaceous; eyes small, oval, in length about as long as the third antennal joint, and situated before the lateral middle line of the head, *i.e.*, the space behind them is greater than in front. Vertex roundly convex. The third antennal joint not quite twice the length of the fourth, which is longer than the second. Head wider than the thorax. Thorax more coarsely punctured than the head; the mesonotum impunctate, the pro—and metapleuræ rufose; the sides of mesothorax rough, very slightly concave; eyes of median segment bluntly serrate. Abdomen about as long as the head and thorax united; the hairs on the first and second segment black. Pygidium apparently finely punctured; ventral segments fringed with fulvous hairs. Legs: femora and coxae piceous; tibiae and tarsi ferruginous; the tibial spines (6 in a row) black; the hair long and pale fulvous.

Length, 9 mm.

—MUTILLA BHUDDA, *sp. nov.*

Black, the thorax ferruginous; an oval spot on the base of second segment and the whole of the third pale golden-fulvous. Head narrower than thorax, very coarsely punctured, almost reticulated; eyes large, oval, antennal tubercles impunctate; vertex not much raised above the eyes, roundly convex; the clypeus fringed with long fulvous, the rest of the head sparsely with fuscous hairs; mandibles piceous in the middle. Scape covered with pale fulvous hairs; the flagellum with a pale down, brownish beneath; the third joint not much longer than the fourth, shorter than the fourth and fifth united. Thorax coarsely longitudinally reticulated, the pleurae entirely impunctate; becoming gradually but not much narrowed from extreme base to apex; the edges rough, but without any distinct tubercles; apex of median segment obliquely sloped. Abdomen longer than the head and thorax united; the first segment dilated, longitudinally punctured on the second segment; the others with their apices shining, impunctate, glabrous; the pygidium coarsely punctured; the extreme apex finely transversely striated; the apical ventral segments fringed with long fulvous hairs. Legs: the femora sparsely covered with longish blackish hairs; the tibiae and tarsi thickly with pale fulvo-golden; the calcaria and the bristles on the underside of the tarsi rufous; the four tibial spines stout black.

Length, 11 mm.

—MUTILLA SERRATULA, *sp. nov.*

Black; the thorax red; the scape and legs pale rufo-testaceous; the apex of second abdominal segment with a pale fulvous band (dilated in the middle). Head as wide as the thorax, coarsely punctured; the antennæ tubercles impunctate. Pale testaceous, as well as the cypleus; mandibles reddish, the apices piceous black; eyes large, oval, reaching close to the top of the head. The third joint of antennæ about one-half longer than the fourth. The sides of the thorax coarsely irregularly serrate, contracted in the middle; closely and coarsely longitudinally punctured; the apex of median segment above ∧-shaped; coarsely serrated; the acute apex terminating in a spine. The sides of the median segment serrate; the pleuræ impunctate. First abdominal segment not dilated; the apical pale, testaceous, impunctate; the apical ventral segments fringed with pale fulvous hairs. Tibiæ and tarsi sparingly covered with testaceous hairs; the apices of femora fuscous.

Length, 5 mm.

—MUTILLA INSULARIS, *sp. nov.*

The antennæ and head black; thorax ferruginous; abdomen black, with two large oval united fasciæ on the second segment; the third segment and the apex of the fourth, golden-fulvous; legs black, the femora for the greater part ferruginous. Head coarsely rugosely punctured; the hairs fulvous. Eyes large, oblong; reaching quite close to the top of the head, which is slightly convex—Antennal tubercles red, shining, finely striated. Antennæ inclining to piceous beneath towards the base; the third joint twice the length of the fourth; the second and third joints subequal. Thorax bluntly rounded at base and apex, twice the length of head, more strongly punctured than the head; the pleural punctured, except at the apex ; the sides of the thorax above rough, almost straight—pygidium longitudinally striated ; the sides fringed with long fulvous hair— apical ventral segment slightly fringed with fulvous hair ; the basal segment with large deep punctures ; the others with the punctures much smaller. Legs covered thickly with long pale fulvous hairs, rufo-fulvous on the tarsi ; the calcaria ; the tibial spines four, pale.

Length, 11 mm.

Mutilla Kauarae Cam.

Head covered all over with a golden fulvous pubescence completely hiding the colour ; the basal three joints of the antennæ and the thorax and legs ferruginous ; abdomen black, segments three and four the apex of the second and the apical covered with golden fulvous pubescence. Head a little wider than the thorax ; eyes large, oval ; mandibles ferruginous at the base ; palpi testaceous ; antennæ stout, the third joint a little shorter than the fourth and fifth united. Thorax about twice as long as the head, rounded at base and apex ; the sides concave, the edge rough, hardly wider at the apex than at the base ; the thorax above coarsely rugosely punctured ; the pluræ impunctate, shining ; apex of medium segment with a gradually rounded slope. Abdomen as long as the head and thorax united, subsessile ; coarsely punctured ; the fourth and fifth segments bearing long, dull, fulvous hairs ; pygidium longitudinally striated, the apex more finely transversely : the sides fringed with long golden hairs ; ventral segments punctured, fringed with fulvous hairs. Legs covered with fulvous hair ; the tibiæ with four long spines ; the coxæ are black ; the hair of tarsi dense and long.

Length 11 mm.

Mutilla Constanceae, sp. nov.

Black, the thorax pale, ferruginous above, an irregular

mark on the base and apex of the second segment and the third segment white. Antennæ with the third joint about one quarter longer than the fourth; the first at apex, the second and the terminal beneath more or less piceous. Head not much wider than the thorax, coarsely punctured, densely covered with silvery hair; the antennal tubercles piceous, aciculate. Eyes oval, moderate, reaching quite close to the top of the head. Thorax quadrate, rounded at base and apex, the sides rough, slightly concave; above coarsely longitudinally punctured, the pleuræ impunctate, densely covered with white hairs, the upper also densely covered with white hairs. Abdomen as long as the head and thorax united, subsessile, dilated at the base of second segment, becoming gradually narrowed to the apex; coarsely punctured, in the middle bearing rufous hairs, the sides with silvery hair; pygidium rufous, longitudinally striated, the ventral segments fringed with long silvery hairs. Legs covered with long silvery hairs; the tibial spines testaceous; the spurs white.

Length, a little over 6 mm.

MUTILLA YERBURGHI, *sp. nov.*

Metallic-blue, covered with a whitish pubescence; the antennæ black; wings fusco-hyaline. Antennæ of moderate length; the third joint shorter than the fourth. Eyes emarginate. Head punctured, behind the ocelli almost smooth; a channel runs down from the lateral ocelli. Thorax rather strongly punctured, very slightly convex; the median segment reticulated; pro- and metapleuræ with an oblique excavation, shining, almost smooth except for an indistinct striation. Parapsidal furrows distinct. Abdomen strongly punctured; the second and following segments thinly fringed with pale hair. Thorax truncated in front, rounded behind. Legs covered with white hairs; the calcaria white. Abdomen subsessile.

Length, 9 mm.

Mutilla pulcherina, *Smith*.

Head and thorax ferruginous, the antennæ, mandibles, legs and abdomen black : a small, somewhat triangular, spot on the basal segment, a broad band, dilated roundly in the middle at the apex, on the base of the second ; the third segment entering a broad band in the fourth (the latter two interrupted in the middle), and a fringe on the apical abdominal segments, golden fulvous. Antennæ short, thick ; the third joint about one-half longer than the fourth, and shorter than the fourth and fifth united. Head narrower than the thorax, very coarsely rugosely punctured ; the antennal tubercles shining, impunctate ; eyes small, oval, vertex convex, not much elevated above the eyes ; the long hair on the vertex blackish ; on front and oral region fulvous. Thorax coarsely longitudinally, irregularly strialate ; but the furrows are not continuous ; mesopleuræ shining, impunctate. The sides become gradually and slightly narrowed from base to apex ; a little before the middle there is a stout tooth ; and there is a blunt tubercle a little beyond the middle. Abdomen distinctly longer than the head and thorax united, subessile ; pygidial area apparently strongly transversely aciculate ; the second ventral segment very coarsely transversely punctured. Legs (including the tarsi) densely covered with long fulvous hair ; the femora coarsely punctured ; the calcaria pale testaceous.

The band on the second abdominal segment varies, and may become broken up into three rounded spots.

M. Aurifex is probably a variety ; but the description is imperfect.

Mutilla interrupta, *var?*

Black ; the thorax rufous ; the abdomen with two oval spots on the base of the second segment, and a broad

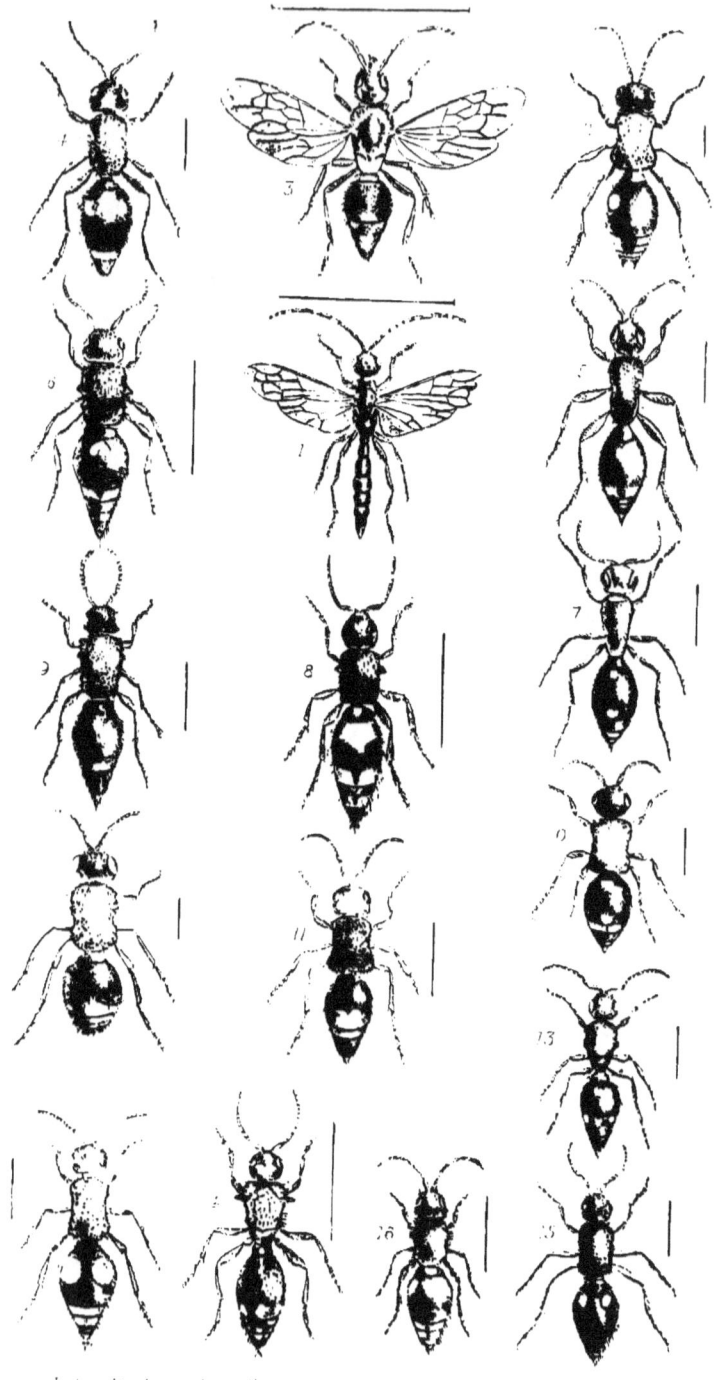

interrupted band on the third and fourth segments, white. Antennæ with the third joint about twice the length of the fourth. Head wider than the thorax, coarsely rugosely punctured; the lower part bearing a white pubescence. Eyes elongate. Thorax quadrate much more strongly and deeply punctured than the head, the pleuræ impunctate; the sides of the thorax above slightly concave, the apex more dilated than the base, the apex of median segment subperpendicular. Abdomen a little longer than the head and thorax united, narrowed at the base; pygidium finely longitudinally striated. Tibial spines pale, four in number. Nearly 9 mm.

APTEROGYNA, *Latr.*

1. APTEROGYNA MUTILLOIDES, Smith, *Cat. Hym.*, III., 64, 5.
   Hab. India.

MYRMOSIDA, *Smith.*

1. MYRMOSIDA, PARADOXA, Smith, *Proc. Linn. Soc.*, II., 88, 1, tab. 2, fig. 1.
   Hab. Singapore.

THYNNIDÆ.

ISWARA *West.*

1. ISWARA LUTENS, Westwood, *Trans. Ent. Soc.*, I. pl. 7, f. 5.
   Hab. India.
2. ISWARA FASCIATA, Smith, *Ann. Mag. Nat. Hist.* 253.
   Hab. Sind.

[*From Volume 41, Part II.*, of "MEMOIRS AND PROCEEDINGS OF THE MANCHESTER LITERARY AND PHILOSOPHICAL SOCIETY," *Session 1896-7.*]

Hymenoptera Orientalia, or Contributions to a Knowledge of the Hymenoptera of the Oriental Zoological Region. Part V.

BY

PETER CAMERON.

MANCHESTER:
36, GEORGE STREET.

1897.

IV. Hymenoptera Orientalia, or Contributions to a knowledge of the Hymenoptera of the Oriental Zoological Region. Part V.

By PETER CAMERON.

[*Communicated by J. Cosmo Melvill, M.A., F.L.S.*]

Received November 9th. Read November 17th, 1896.

In view of the fact that Colonel C. T. Bingham is at present engaged on a Monograph of the Indian Hymenoptera, I deem it advisable to give in this part of my paper descriptions of new species only, leaving the information I possess regarding the distribution and habits of the known species to be dealt with in another paper.

Compared with the immense number of parasitic Hymenoptera (Ichneumonidæ, Braconidæ, &c.) known from the Nearctic and Palæarctic zoological regions they are but feebly represented, in fact they are almost absent in the southern parts of the Indian Peninsula; but they appear to be more numerous in Ceylon, and are probably not uncommon in the Himalayas.

ICHNEUMONIDÆ.

ICHNEUMON CLOTHO, *sp. nov.* (Pl. 3, f. 1).

Niger, abdomine cæruleo; linea antennarum, orbitis oculorum, scutello, post-scutelloque albis; alis fusco-hyalinis.
♀. Long. 15 mm.

Hab. Mussouri (*Rothney*).

Head black, shining, the face strongly, the front and vertex much less strongly punctured; the orbits on the top and bottom and on the inner side, on the bottom very broadly; on the sides narrowly, and dilated in the middle, especially at the top, yellow; the mandibles thickly covered with dull fulvous hair; rufous before the apex. Antennæ stout; 7-8 joints near the middle on the underside yellowish-white; the scape punctured; on the underside covered with pale golden hair. Thorax black; the edge of the pronotum, scutellum and post-scutellum yellow, the mesonotum closely punctured; thickly covered with a short, dull fulvous pubescence; the scutellum punctured: the post-scutellum hardly so; the sides of the former smooth, shining, and impunctate; the depression at the side of the post-scutellum also smooth and shining and with a distinct margin; the median segment coarsely irregularly striolated at the base; the middle of the apex closely transversely striated; the sides much more strongly transversely striated; all the areæ clearly defined. The lower part of the propleuræ irregularly obliquely striated. The mesopleuræ closely punctured; the middle with some irregular longitudinal striations; the metapleuræ closely punctured, running into irregular striations at the apex. Legs covered with pale pubescence; the outer and lower half of the four anterior coxæ, the hinder coxæ entirely black, except a yellow spot in the middle above; the front femora are rufo-testaceous on the underside, the middle pair towards the apex only; the front tibiæ testaceous in front; the middle pair rufo-testaceous towards the apex; the hinder pair quite black and both the hinder pairs have a broad white mark near the base; the four anterior tarsi dull testaceous, the hinder pair dark, with the third and fourth joints white. Abdomen shining, bright blue, except the petiole, which is darker; shining, almost glabrous; the petiole at the apex before the gastrocœli with a strongly punctured spot; the gastrocœli shallow, large, the base with some striations; the apex flesh-coloured.

ICHNEUMON TAPROBANÆ, *sp. nov.*

Niger, pedibus flavis: coxis trochanteribusque nigris, alis fulvo-hyalinis, stigmate fusco. ♂. Long. 11 mm.

Hab. Ceylon (*Rothney*).

Head bearing a white down; closely punctured, the clypeus with the punctures widely separated; the palpi testaceous. Thorax closely punctured; the depression at the base of the scutellum wide, deep; the scutellum finely punctured, shining; the centre fuscous; its apex finely longitudinally striated; the post scutellum finely rugose; the median segment rugose; there are two straight keels at the base, forming an almost square area; the lateral keels end in stout, blunt spines; the apex oblique, with two straight keels down the centre. The propleuræ finely longitudinally striated at the base; the mesopleuræ punctured, an oblique band of fine striations in the middle; the metapleuræ longitudinally striated, indistinctly at the base, much stronger at the apex. The front coxæ on the lower side are thickly covered with fulvous hair; the middle femora are fuscous on the underside at the base. At the top the areolet is narrowed, being there as long as the space bounded by the second transverse cubital and the recurrent nervures. Abdomen black; the apex of the petiole finely rugose, keeled down the sides; the gastrocœli finely striated at the base.

ICHNEUMON INTAMINATUS, *sp. nov.*

Niger; femoribus posticis rufis: alis hyalinis, nervis stigmateque nigris. ♂. Long. 9 mm.

Hab. Mussouri (*Rothney*).

Head entirely black; the front and vertex very shining, sparsely covered with white hair; the face and clypeus almost impunctate, covered with long white hair; the mandibles and palpi black. Antennæ as long as the body, black, the scape with longish pale hair. Thorax shining, the pleuræ and sternum with long white hair: the median segment with an oblique slope at the apex; its base smooth, the rest rough, the apex with a few irregular striations; the apical part of the propleuræ near the middle with a few striations: the rest of the pleuræ obscurely punctured; the mesopleuræ and the metapleuræ at the apex on the lower side striolated. Legs thickly covered with short white hair; black, including the calcaria; the hinder femora red. The areolet is a little narrowed at the top, being there as wide as the space bounded by the first transverse cubital and the recurrent nervures. Petiole coarsely aciculated, keeled down the centre and the sides; the apex smooth, shining; the other segments smooth and shining.

ICHNEUMON VACILLANS, sp. nov.

Niger: tibiis tarsisque anterioribus, femoribus basique tibiarum posticarum, rufis; alis hyalinis, nervis fuscis. ♂. Long. 11 mm.

Hab. Mussouri (*Rothney*).

Head black, thickly covered with moderately long white hair, closely but not strongly punctured; the palpi black. Antennæ entirely black; the scape with a longish, the flagellum thickly with a microscopic pile. Thorax entirely black; thickly covered with short fuscous pubescence. Metapleuræ much more strongly punctured than the mesonotum and more opaque; the apex with a gradually rounded slope; obscurely striolated in the middle; a shallow furrow leads down from the spiracles; the pleuræ uniformly punctured. Wings hyaline; the nervures at the base and the stigma in the middle testaceous. Legs: the apical third of the anterior femora, the tibiæ and tarsi, the middle legs except the coxæ and trochanters, the hinder femora except the knees, and the basal third of the tibiæ, red; the calcaria pale rufous. Abdomen black, shining, impunctate.

ICHNEUMON APPROPINQUANS, *sp. nov.*

Niger; multo albo-maculato; pedibus rufis, basi nigris; alis fulvo-hyalinis, stigmate fulvo. ♀. Long. 15 mm.

Hab. Mussouri (*Rothney*).

Head black; strongly punctured; the vertex transversely striolated; the inner orbits from the clypeus to the hinder ocelli broadly yellow, narrowed towards the eyes; at the outer side at the foot of the eyes is a large yellow mark; and on each side of the clypeus is a crescent-shaped mark: palpi pale fulvous. Antennæ black, the upper side beyond the middle broadly white. Thorax closely punctured, covered with a pale short down; the edge of the pronotum, tegulæ, tubercles, scutellum, post-scutellum, two large marks on the median segment at the tubercles; and a large mark on the mesopleura, clear yellow. The apex of the pro- and meso-pleuræ crenulated; on the median segment are only the basal areæ; the apex oblique; at its top are some stout longitudinal keels; the centre coarsely rugosely punctured; the sides with stout transverse keels. The four front coxæ are broadly white at the base in front; the hinder pair have a large white mark at the base above; the trochanters black; the four anterior white at the apex; the hinder with a small white mark above; the apex of the fore femora, the tibiæ and the base of the tarsi white in front; the tarsi black, paler at the base. Wings fulvo-hyaline, the nervures fuscous, paler towards the apex; the stigma fulvous; the areolet at the top nearly as long as the space bounded by the recurrent and second transverse cubital nervures. Abdomen black; the apex of the petiole and a large mark on each side of all the other segments, white. Petiole smooth, shining, longitudinally striolated in the middle before the white; gastrocœli broad; their base roughly and irregularly striolated.

ICHNEUMON HYPOCRITA, sp. nov.

Niger, annulo antennarum, facie, scutello, coxis, trochanteribus tarsisque posticis albis; pedibus fulvis; trochanteribus posticis tarsisque posticis late nigris; alis fusco-hyalinis. ♂. Long. 10 mm.

Hab. Mussouri (Rothney).

Head black, closely punctured; the face thickly covered with short white hair; below the antennæ entirely white, including the oral organs. Antennæ black, the scape beneath, and a broad belt (9 joints) beyond the middle, white. Thorax black; the tegulæ, tubercles, scutellum, and post-scutellum white; closely punctured; the scutellum with widely separated punctures; at the sides of the post-scutellum is a row of stout keels; the areæ on the median segment clearly defined; the supra median longer than broad, bulging out at the sides; the lateral wider than it and impunctate; the rest of the segment transversely striated. Pleuræ strongly punctured; the lower part of the propleuræ strongly striated; the metapleuræ rugosely punctured. The four anterior coxæ and trochanters white; the rest of them fulvous; the hind coxæ black; the hind femora black except at the extreme base; the hind tibiæ broadly black at the apex; the hind tarsi white except at the base. Wings hyaline, with a faint fuscous tinge; the stigma and nervures black; the areolet a little longer than broad; the transverse cubital nervures straight, parallel; the recurrent nervure is received shortly beyond the middle. Abdomen black, shining, smooth; the second and third segments at the apex fulvous.

ICHNEUMON ROTHNEYI, sp. nov. (Pl. 3, f. 2).

Niger; abdominis segmentis 2—6 ferrugineis; tibiis tarsisque sordide testaceis; alis flavo-hyalinis, stigmate fulvo. ♂. Long. 17—18 mm.

Hab. Mussouri (Rothney).

Head black, strongly and closely punctured; thickly covered with short grey hair; the mandibles piceous before the apex. Antennæ black, 7—8 of the middle joints yellowish-white. Thorax black, closely punctured; a line on the pronotum and the scutellum yellow; the latter smooth and almost impunctate; its apex black. The median segment rugose; thickly covered with short white hair, its apex hollowed in the centre; its sides with stout keels: there are two central keels widely separated; the space between these and the bordering keel stoutly transversely striated, it being also striolated obscurely on the innerside of the central keel; the basal area is a little wider than long. All the coxæ and trochanters are black and thickly covered with a pale down; the anterior femora slightly; the middle pair even more so and the hinder pair still more so, the apex of the hinder tibiæ and the apical joints of the tarsi, blackish. Wings hyaline, but with a distinct fulvous tinge; the stigma clear fulvous; the nervures fuscous, darker towards the base; the top of the areolet is a little less in length than the space bounded by the second transverse cubital and the recurrent nervures. Abdomen ferruginous; the petiole and the base of the second segment, black; the petiole smooth at the base; the sides at the apex coarsely punctured; the raised central part closely longitudinally striated. Gastrocœli large, deep, with a few stout, irregular keels; the second segment between longitudinally striated; the apical segments are smooth and impunctate.

ICHNEUMON BUDDHA, *sp. nov*.

Niger, scutello flavo; flagello antennarum medio late albo; pedibus rufis; apice tibiarum posticarum tarsisque posticis nigris; abdomine late rufo, apice albo. ♀. Long. 8 mm.

Hab. Mussouri (*Rothney*).

Head black, the sides of the clypeus and the greater part of the mandibles in middle rufous; the eyes very distinctly margined on the inner side; the front and vertex closely punctured; the clypeus, except at the base, with very few punctures. Antennæ stout, sparsely and shortly pilose; the middle of the flagellum broadly white beneath and at the sides. Thorax black; the pronotum yellowish in the middle; the mesonotum coarsely alutaceous; the scutellum shining; post-scutellum narrowed towards the base; the part at the sides of the scutellum behind with strong, stout keels, which become wider from the inner to the outer side; the part at the sides of the post-scutellum crenulated. Median segment large; the apex with an oblique slope, roundly excavated in the centre; coarsely punctured; the sides at the apex irregularly obliquely striated; all the areæ completely defined; the central coarsely transversely striated; there are two stout keels bordering the central depression on the apical part. Propleuræ at the apex strongly striolated; the part above this coarsely punctured, the puncturing becoming finer towards the top; above the front coxæ are two stout, sharp keels, forming between them a sharp channel. All the coxæ and base of trochanters black; the apex of the hind tibiæ and the tarsi black: the middle tarsi fuscous. The basal half of the petiole black; the apical rufous: coarsely punctured: and with two sharp keels down the middle. The second and third segments entirely rufous: the apical two cream white above, except at the side of the penultimate: the ventral surface broadly rufous at the base; gastrocœli striated at the base.

ICHNEUMON VISHNU, *sp. nov.*

Long. fere 10 mm. ♂.

Hab. Mussouri (*Rothney*).

A species very like the above described species. It

seems to differ from it in too many points of colouration and structure to be its ♂. For example, the supra-median area on the median segment here is rounder and completely separated from the base, which is not the case with the other species; it is, further, fully larger, while if it were the ♂, it should be smaller.

Antennæ as long as the body, the scape covered with long fuscous hair; a white band of six joints beyond the middle. Head black; strongly punctured; the clypeus with only a few punctures on its apical margin. The orbits, except at the top, yellow, the inner band being the wider, especially at the bottom: the mandibles broadly yellow in the middle; the labrum fringed densely with golden hair. Thorax black: the edge of the pronotum, but not completely, the tegulæ, tubercles, and scutellum broadly at the sides, but not uniting at the apex, a line at the sides in front of it, and the post-scutellum, yellow. Pro- and meso-thorax closely punctured; the scutellum thickly covered with brownish hair; at the sides of the post-scutellum there are some stout keels. The median segment has an oblique slope at the apex; the central area complete, rounded at the base, almost transverse at the apex; the other nervures are not so clearly defined; the apex is transversely striated. Legs rufous; the coxæ, trochanters, apex of the hinder tibiæ and the posterior tarsi, black; the fore coxæ yellowish white at the apex. The areolet is much narrowed at the top: the transverse cubital nervures almost meeting there. Petiole black, a yellow band across its apex; the second and third segments rufous; the apical two segments white above; beneath the 2—4th segments are rufous.

ICHNEUMON CONFUSANEUS, *sp. nov.*

Niger, flagello antennarum albo; tibiis tarsisque anticis testaceis; abdomine late rufo; alis hyalinis, stigmate fusco. ♀. Long. 6 mm.

Hab. Trincomali, Ceylon (*Yerbury*).

Antennæ black, the 8—14th joints clear white beneath. Head closely punctured all over; the clypeus with the punctures more widely separated; the mandibles before the teeth rufous. Pro- and meso-notum closely and rather strongly punctured; the scutellum with the punctures smaller and more widely separated. Median segment with the areas complete; coarsely punctured; the apex in the middle closely transversely striated. The apex of the propleuræ strongly longitudinally striolated; on the mesopleuræ there is a shining impunctate spot below the hind wings. The middle tibiæ are darker coloured than the anterior, especially in front. The areolet is narrowed at the top, being there not much wider than the space bounded by the recurrent and the transverse cubital nervures; the recurrent being received nearly in the middle of the cellule. Petiole shining, black; the apex, except in the middle, with some distinctly separated punctures; the second, third, and base of the fourth segment ferruginous; gastrocœli smooth.

Ichneumon inquietus, *sp. nov.*

Long. 6 mm.

Hab. Trincomali, Ceylon (*Yerbury*).

Is nearly related to *I. confusaneus*; but differs in having the legs without black, except on the apex of the hinder femora and tibiæ; the propleuræ more completely and strongly striolated, and the second abdominal segment closely longitudinally striated.

Head closely punctured, immediately over the antennæ obscurely striated transversely; the clypeus more shining, with the punctures much more widely separated; the mandibles pale yellow, piceous before the apex. Antennæ black; the 9—12th joints clear white. Mesonotum closely punctured; the scutellum shining; sparsely punctured, especially at the base. Median segment coarsely punctured ; the apex in the middle transverely striolated. The mesonotum punctured ; the pro- and meta-pleuræ on the lower part longitudinally striolated. The four anterior legs fulvous; the coxæ and trochanters yellow; the hinder legs have a more reddish tint, including the coxæ and trochanters ; a spot on the underside of the coxæ, the apex of the femora and the tibiæ black. Petiole black, shining ; the second and third segment rufous ; the base of the second depressed and with an interrupted transverse smooth shallow furrow at the apex of the depression ; strongly longitudinally striolated, the striæ becoming very faint towards the apex, especially in the middle, which is aciculated.

ICHNEUMON INTEGRATUS, *sp. nov.*

Niger, facie, scutello, maculis 2 metanoti abdominisque segmentis late, flavis ; alis fulvo-fumatis, stigmate fulvo. Long. 15 mm.

Hab. Mussouri (*Rothney*).

Head black, densely covered with white, behind with fulvous hair ; the face below the antennæ yellow ; fulvous in the middle (perhaps through discoloration) ; the apex of the clypeus black ; the mandibles reddish towards the apex ; the base densely covered with fulvous hair ; the palpi fulvous. Thorax thickly covered with pale pubescence ; the tubercles, tegulæ, a short line in front of them, scutellum, a line on the post-scutellum, and two triangular marks on the median segment near the spiracles, yellow. Propleuræ irregularly striated towards the apex below. The median segment has only the basal area clearly defined ; it is broader than long, rough ; the centre with four stout keels, converging towards the apex ; the apex has an abrupt slope ; the central area closely and roughly transversely striated ; the lateral areæ with the transverse keels much stouter, more widely separated, and they are divided into two by a stouter keel. Coxæ black; the four hinder broadly yellow at the base; the apical joint of the four anterior trochanters pale; of the hinder pair entirely rufous; the apex of the hinder femora and of the hinder tibiæ black; the hair is fulvous on the coxæ. The alar nervures fuscous, pale at the base: at the top the areolet is as wide as the space bounded by the second transverse cubital and the recurrent nervure. Petiole shining at the base: the apex closely longitudinally striolated; the gastrocœli shining; the base with some stout keels; the yellow on the second and third segments, broad; the apical segments entirely rufous, the fourth and fifth only rufous at the base.

ICHNEUMON NUMERICUS, *sp. nov.*

Long. 15 mm. ♂.
Hab. Mussouri (*Rothney*).

Resembles *I. integratus;* but has the yellow markings much more expanded on the thorax; the pleuræ having two large marks; the apex of the median segment yellow, and the antennæ yellow, broadly black at the apex.

Head rather strongly punctured, the lower part densely covered with white, above the antennæ with longer fuscous hair; below the antennæ entirely yellow; the inner orbits broadly to the hinder ocelli, and the outer to near the level of the lower, this latter belt becoming gradually narrowed to the top, yellow. The front in the middle transversely striated: the mandibles and palpi yellow: the former piceous at the apex: palpi yellow. Antennæ as long as the body, rufous-yellow, the apex broadly black; the three basal joints black above. Thorax closely punctured; thickly covered with short hair, which is darker on the mesonotum, whiter on the rest; a broad band on the pronotum; the scutellum, post-scutellum, the apex of the median segment, except a small black mark in the middle at the apex, a triangular mark on the lower part of the propleuræ, a large mark on the lower side of the mesopleuræ, and an oblique one on the metapleura, yellow. On the median segment only the basal area is defined; it is longer than broad, smooth; the apex with four stout longitudinal keels, its extreme apex, yellow; the sides at the base rugosely punctured; the apex transversely striated, more strongly at the sides; the central keels stout, straight. The four front coxæ entirely yellow; the hinder black; the apex broadly rufous; the four front legs entirely rufous, yellower in front; the hinder rufous; the apex of the tarsi broadly black. Petiole black; the sides margined; stoutly keeled at the apex, which is strongly longitudinally striolated, raised at the middle, depressed at the sides; the base of the second segment striated; the gastrocœli shallow; the black bands on the second and third segments broad, triangularly produced in the middle at the base; in the centre of the fourth segment at the apex is a black mark, triangularly produced at the base.

ICHNEUMON AGRAENSIS, *sp. nov.*

Fulvus, pedibus posticis nigro-maculatis; alis fulvo-hyalinis, stigmate fusco. ♂. Long. 13 mm.

Hab. Agra (*Rothney*).

Head luteous, the orbits paler; covered with a white microscopic pubescence; the face closely covered with shallow punctures; the apex of the clypeus rounded; the tips of the mandibles blackish. Scape of antennæ luteous; the flagellum, brownish beneath, darker above. Mesonotum of a darker tint than the rest of the body; rough in texture; the scutellum with large punctures; and covered with long fuscous hair, large, raised above the level of the mesonotum; a deep depression at its base; its apex oblique. Median segment closely rugosely punctured; the base with the punctures larger and more widely separated; the extreme base impunctate. There is a central pear-shaped area, and two wide lateral ones; the apex is rounded and transversely striolated in the middle. The lower part of the propleuræ shining, impunctate: the upper with shallow punctures, the meso- and metapleuræ closely punctured; an impunctate spot on the mesopleuræ near its apex. Legs fulvous; the apex of the hind femora, of the hinder tibiæ and the hinder tarsi except at the base, black. The areolet at the top is as wide as the space bounded by the recurrent and the second transverse cubital nervures. Except at the base the abdomen is closely punctured, the apex of the petiole, depressed at the sides; the gastrocœli large, the innerside at the base striolated, the outer punctured, the space between longitudinally striated; the sixth joint entirely, the seventh, black with a large white mark in the middle at the apex; the two apical segments entirely black beneath.

CRYPTUS INFERNALIS, sp. nov.

Ferrugineus, capite, antennis, abdominisque apice late nigris; flagello antennarum annulo late albo; alis hyalinis, nervis fuscis. Long. 7 mm.

Hab. Agra (*Rothney*).

Head black: the orbits from the top of the frontal depression to the occiput, white, the white mark narrowed at base and apex; the frontal depression transversely striated; the palpi testaceous. Antennæ black, the middle of the flagellum broadly white. Thorax entirely red; the pro-mesonotum and scutellum shining; almost impunctate; the depression at the base of the scutellum crenulated. Median segment closely rugosely punctured; the base laterally shining and impunctate; at the base and at the top of the flat part is a transverse keel which bulges backwards in the middle, the basal one being rounded, the apical transverse at the base; the apex has an oblique slope; the pleuræ closely punctured; the propleuræ at the base more shining and obscurely striolated. The four anterior legs rufous; the hinder femora and the coxæ black, except at the base and apex; the hinder tibiæ and tarsi black, the former only black behind. The petiole is broadly black at the base; smooth and shining, the apex without keels and not raised in the centre at the apex; its apex and the second segment ferruginous; the other segments black except the last, which is milk-white above. Gastrocœli absent. Areolet almost square.

CRYPTUS INDICUS, sp. nov.

Niger, albo-maculatus; pedibus anterioribus pallidis; coxis, trochanteribus femoribusque posticis rufis; tibiis tarsisque posticis nigris, basi albis; alis hyalinis, apice fumatis. ♂.
Long. 8—9 mm.

Hab. Mussouri (*Rothney*).

Head shining, the front sparsely punctured; below the antennæ, including the oral organs and the inner orbits to the ocelli broadly, white; the tips of the mandibles black; the basal portion of the antennæ white beneath; the apical brownish. Thorax black, shining, minutely punctured; the prothorax in front, a curved mark narrowest on the outerside, on the side of the mesonotum at the base, the tubercles, tegulæ, scutellum, and a mark on the apex of the metapleuræ over the coxæ, white. Pro- and meso-notum punctured; the scutellum impunctate, the median segment much more strongly punctured and without any keels. The four front legs are entirely pallid yellow; the hind coxæ, trochanters, and femora red; the apex of the hind femora, tibiæ, and tarsi, black; the base of the hind tibiæ and the greater part of the metatarsus at the base, white. Wings clear hyaline, the apices of both wings smoky; the nervures fuscous; the areolet shortly appendiculated at the top; the recurrent nervure received in the basal third of the cellule. Abdomen very smooth, shining; the petiole entirely white on the basal half; its apex narrowly, the base of the second segment, its apex narrowly, the base and apex of the third and fourth, broadly, white; the ventral surface for the greater part white.

CRYPTUS ORIENTALIS, *sp. nov.*

Niger, pedibus abdomineque late rufis; alis fusco-hyalinis, nervis testaceis. ♀. Long. 12; terebra 4 mm.

Hab. Mussouri (*Rothney*).

Head black; the apex of the clypeus, the orbits narrrowly, except near the top behind, the base of the mandibles, a line at their base joined to the eyes, pale testaceous; the palpi fuscous, testaceous at the base. Antennæ black; the 6—8 joints pale testaceous beneath. Thorax black; closely punctured; the lower half of the propleuræ strongly longitudinally striolated; the parapsidal furrows complete, deep, broadest at the base; the scutellum closely punctured; the post-scutellum shining, and bearing a few scattered punctures; on the base of the median segment are two large areas, curved; truncated at the sides, the space enclosed being finely rugose; between the basal and the apical keels the front is strongly irregularly striolated; the central keels being the larger and most regular; the spines large, somewhat triangular. The apex has an oblique slope; the centre coarsely coriaceous; the sides with stout transverse striations. Coxæ and trochanters black; the anterior trochanters testaceous at the apex; the hinder tibiæ infuscated, especially towards the apex; the hind tarsi rufo-testaceous, the metatarsus except at the apex, and the apex of the terminal joint black. At the top the areolet is as wide as the space bounded by the second transverse cubital and the recurrent nervure. Abdomen shining, impunctate; bare; black; the apex of the petiole, the apex and sides of the second segment, the third segment except at the base and the others almost entirely, rufous.

HEMITELES VEDA, *sp. nov.*

Ferrugineus, thorace nigro-maculato; alis fulvis. Long. 15; terebra 4 mm.

Hab. Trincomali, Ceylon (*Yerbury*).

Head ferruginous, the part above the antennæ, and a triangular mark leading down to it from the ocelli, black, the part enclosing the ocelli being also black; strongly punctured, the clypeus and the part immediately over the antennæ, smooth; the inner orbits below the antennæ obscure yellow; the inner orbits above the antennæ distinctly margined; the clypeus near the base of the mandibles, black. Antennæ bare; from the thirteenth joint brownish beneath, blackish above. Thorax rufous; a small mark on the propleuræ, the mesopleuræ broadly at the base, narrowly at the top and down the apex, the metapleuræ except a mark in the centre leading into a smaller one at the side, black; the extreme base of the median segment, its apex and two oblique marks there, black; the metanotal keels almost obsolete at the base; towards the apex there are two straight central and an oblique lateral fairly well indicated; the lower side of the propleuræ obliquely striolated; the base and apex of the mesopleuræ narrowly longitudinally striated; the base of the metapleuræ crenulated, and on the lower side there is a stout curved keel. Legs ferruginous, the tips of the tarsi and a large mark on the hinder side of the posterior coxæ, black. Wings fulvous, lighter coloured at the apex; the stigma and costa fulvous; the nervures blackish; the areolet wider than long, narrower at the bottom than at the top through the first transverse cubital nervure being sharply, the second slightly, oblique; the recurrent nervure is received in the basal third of the cellule. Petiole shining, the apex finely punctured, and with an elongated depression; the base broadly black; the rest of the abdomen shagreened.

MESOSTENUS HIMALAYENSIS, *sp. nov.*

Niger, albo-maculatus; pedibus fulvis; coxis anterioribus albis, basi tibiarum tarsorumque posticorum late nigris; alis hyalinis; nervis fuscis. ♂. Long. 9 mm.

Hab. Himalayas.

Antennæ as long as the body; black; the scape beneath and a broad band beyond the middle, white. Head shining; the face closely punctured; the front obscurely striolated; below the antennæ, the oral region except the apices of the mandibles, the orbits except near the top of the eyes, white. Thorax black; the mesonotum closely punctured; a broad line on the pronotum, tegulæ, tubercles, a mark on the centre of the mesonotum, scutellum, post-scutellum, three marks on the median segment in a triangle, a mark at the base of the mesopleura, a smaller one at the apex nearer the breast, a somewhat triangular mark below the hind wings, a large pear-shaped mark on the metapleura and the greater part of the mesosternum, white. Median segment with a gradually rounded slope, coarsely punctured; the basal white mark is longer and narrower than the apical. The four front coxæ and trochanters white; the hinder red like the femora; the basal joint of the trochanters blackish; the apex of the second, the third, and the fourth tarsal joints are white. Wings hyaline, the areolet quadrangular; the recurrent nervure received at its apex. Abdomen black, shining, impunctate; all the segments broadly white at the base above and beneath.

ROTHNEYIA, *gen. nov.*

Differs from all known Ichneumonidæ by having only three visible abdominal segments, the third ending at the apex in a semicircle which forms at each side a stout tooth; the scutellum projects at each side in a stout triangular tooth; there are two large spines on the centre of the median segment at the side. Antennæ 25-jointed. Legs and wings as in *Ichneumon*.

This genus does not fit well into any of the subtribes of *Ichneumonides*. The alar neuration is quite as in *Ichneumon*; but otherwise the genus differs completely; and, as regards the abdomen, it can only be compared with some Braconidæ such as *Chelonus*. The form of the spiracles I cannot determine from the roughness of the median segment.

ROTHNEYIA WROUGHTONI, *sp. nov.* (Pl. 3, f. 3).

Nigra, petiolo ferrugineo; pedibus rufis; geniculis, tibiis tarsisque posticis, nigris; alis hyalinis, basi antennarum late rufis. ♀. Long. 5 mm.

Hab. Mussouri (*Rothney*).

Antennæ black; the 5—6 basal joints of the flagellum brownish; closely covered with a microscopic down, the scape with white hair. Head black; below the antennæ thickly covered with long white hair; the front and vertex punctured, more sparsely covered with fuscous hair. Mandibles depressed at the base; piceous in the middle; the palpi white. Thorax black; the mesonotum more strongly in the centre, which is broadly raised; the scutellum rugosely punctured; the sides raised; the apex between the teeth depressed; the apex of the teeth rufous. The middle of the metanotum between the teeth stoutly bordered or margined all round; the top longitudinally, the apex irregularly transversely striolated; with a semicircular keel at the extreme apex. Pleuræ shining; the lower part strongly transversely striolated; the mesopleuræ at the top punctured, on the lower part more closely punctured; the central part impunctate and with a few striations. Wings hyaline, the nervures fuscous; the cubitus a little narrowed at the top, being there as wide as the space bounded by the second transverse cubital and the recurrent nervures, which, as is also the second transverse cubital, are widely bullated. Legs rufo-testaceous: the apex of the hinder femora and the hinder tibiæ and tarsi, black. Abdomen black; the petiole rufous; covered closely with short white hair; closely and strongly punctured; petiole with the sides strongly keeled; the keels at the dilated apex being continued slightly obliquely to the apex down the middle; the genital armature white.

PIMPLIDES.

PIMPLA PULCHRIMACULATA, sp. nov.

Nigra, late flavo-maculata; pedibus fulvis; alis hyalinis, apice violaceo-maculatis. ♀. Long. 14 mm.

Hab. Trincomali, Ceylon (*Yerbury*).

Head smooth, shining; yellow; the ocellar region, a band leading down from it to a broad transverse band over the antennæ, and the occiput broadly, black. Antennæ nearly as long as the body, black. Palpi testaceous. Pronotum narrowly edged with yellow; the mesonotum black, with two lines in the middle running from the base to the tegulæ, becoming gradually narrower as they do so: the scutellum, except at the apex; post-scutellum and two broad curved lines on the sides of the median segment; the base of the propleuræ, a large mark on the mesopleuræ much narrowed on the lower side, the tubercles, a mark before the middle coxæ, and the metapleuræ, except a black oblique line leading to the spiracles, yellow. A broad black mark, narrowed in the middle down the centre of the median segment. Legs fulvous, the coxæ yellow; a large mark in front of the hind pair and a smaller mark behind, joined together by a broad band at the top, black. Petiole smooth and shining; a broad band in the middle ending before the apex in a large semicircle; the other segments closely punctured; the terminal segments are brownish; the oblique depression on the 2—4th segments distinct; the second segment broadly depressed at the sides at the base, the segment at the outer side of the depression being yellow. The outer half of the cubitus curved; the areolet oblique, shortly appendiculated at the top; the cloud at the apex extends from the costa to about the same distance below the cubital nervure.

PIMPLA TAPROBANÆ, *sp. nov.*

Nigra, pedibus flavis, coxis trochanteribusque nigris; alis fulvo-hyalinis. ♀. Long. 13 mm.

Hab. Ceylon.

Head closely punctured, covered with a short white pubescence; the face projecting, at top forming almost a triangle; clypeus forming a semicircle at the top, where it is obscurely punctured; the apex almost perpendicular; the labrum piceous, fringed with long golden hair; palpi and mandibles entirely black. Pro- and meso-notum thickly covered with fuscous hair; obscurely shagreened; a large square spot on the scutellum and a long one on the post-scutellum, yellow; the median segment broadly raised in the middle at the base; the centre raised; strongly, the sides finely transversely, striated; the centre at the apex with an oblique slope; the sides rather acute at the top. Pro- and meso-pleuræ shining, impunctate, thickly covered with short whitish pubescence: and having a plumbeous tinge. Legs almost bare; the fore trochanters beneath and at the apex all round, yellow. Wings fulvo-hyaline; the stigma testaceous in the middle;

the tegulæ black. Abdomen entirely black: shining, impunctate; the petiole at the base depressed in the middle; oblique; its top somewhat triangularly, its sides much more widely depressed; gastrocœli oblique, smooth, raised in the centre; and from them an oblique furrow leads to the apex of the segment; the oblique furrows on the third segment moderately deep and wide; on the fourth they are shallower.

PIMPLA LAOTHOE, sp. nov.

Nigra, pedibus rufis; coxis, trochanteribus, apice tibiarum posticarum tarsisque posticis, nigris; alis fulvo-hyalinis; nervis fuscis, stigmate fulvo. ♀. Long. 13 mm.; terebra 3—5 mm.

Hab. Mussouri (*Rothney*).

Head black, thickly covered with pale fulvous hair, especially long and thick below the antennæ, where there is in the centre a shining, impunctate line; the front broadly but not deeply depressed; the front ocellus surrounded by a furrow, which is continued down the front to the antennæ; the front with the punctures shallow, especially towards the eyes. Thorax entirely black: the pro- and meso-notum closely punctured, thickly covered with short fuscous hair; scutellum shining, smooth; the punctures shallow, widely separated, the sides much more strongly and closely punctured, except at the base; post-scutellum rugosely punctured; the median segment with a gradually rounded slope, rugose; the centre transversely striated. Propleuræ at the bottom longitudinally striated; at the top are two stout longitudinal keels. All the coxæ and trochanters black; the coxæ beneath thickly covered with fulvous hair, as are also the tibiæ and tarsi; the hinder tarsi black. Abdomen entirely black: above closely, strongly, and uniformly punctured, except at the apices of the segments, which are smooth and shining; the base of the petiole widely depressed, smooth and almost impunctate.

PIMPLA NEPE, *sp. nov.* (Pl. 3, f. 4).

Long. 13 mm.; terebra 4 mm.

Hab. Mussouri (*Rothney*).

Almost identical in coloration with *P. laothoe*, but may be known from it by the absence of the furrow on the front and of the keels on the propleuræ; by the scutellum being more closely and strongly punctured and pale yellow in the centre.

Head closely and strongly punctured below the antennæ; the front widely depressed, impunctate, shining; the palpi dirty testaceous. Pro- and meso-notum strongly and closely punctured, thickly covered with short pale hair; the scutellum thickly punctured behind, more sparsely in front; the top with a pale orange mark; the post-scutellum strongly longitudinally striolated. Median segment with a gradually rounded slope from the base to the apex; coarsely rugosely punctured; the centre broadly raised in the middle towards the apex. Pleuræ and sternum punctured. All the coxæ and trochanters are black; the former on the lower side thickly covered with fulvous hair; the hinder tarsi black, except at the apex. The petiole with a deep impunctate excavation at the base; the other segments closely and somewhat strongly punctured, except at the extreme apex.

OPHIONIDES.

ENICOSPILUS CEYLONICUS, *sp. nov.*

Flavus; alis hyalinis, stigmate fulvo. ♀. Long. 15 mm.

Hab. Ceylon, Trincomali (*Yerbury*).

Antennæ longer than the body, uniformly fulvous; the scape bare, the flagellum with a close microscopic pile. Head fulvous, the face paler, more yellowish; the tips of the mandibles black; the palpi testaceous; the ocelli very large, raised above the level of the eyes, which the hinder almost touch. Mesonotum shining; the scutellum pallid yellow; the base of the median segment depressed in the middle; a stout transverse keel behind it; behind this keel the segment is coarsely shagreened and with an indistinct furrow down the centre; the pleuræ coarsely shagreened. Legs uniformly fulvous, almost bare; wings clear hyaline; the stigma fulvous; the clear bare space contains one large horny mark, with a distinct dark border; above it is a curved spot, and behind two smaller spots. Abdomen darker towards the apex. The cubital nervure is much thickened at the base.

Enicospilus, or *Henicospilus* as the purists would have it, differs from *Ophion* proper in the fore wings having a clear space, which usually contains one or more horny points; and, the stump of the cubital nervure, found well developed in *Ophion*, is absent. In some cases the smooth space is present without having horny points in it, or they are very faint. In either case I believe it will be found that the base of the cubital nervure is thickened, which is not the case with *Ophion, sensu str.* In view of the great similarity of the species of *Ophion*, it seems to me desirable to adopt *Enicospilus* as a distinct genus. Species belonging to it are found in all parts of the world.

ANOMALON DECORUM, sp. nov.

Nigrum, facie, orbitis oculorum, ore, palpis, linea pronoti, tegulis, scutello, coxis trochanteribusque, flavis; alis hyalinis.
♀. Long. 10 mm.; terebra 3 mm.

Hab. Trincomali, Ceylon (*Yerbury*).

The scape yellowish beneath; the flagellum absent. Head shining, sparsely covered with white hair; yellow; the centre of the vertex broadly (the black narrowed towards the bottom), and the occiput, except at the edges, black; the tips of the mandibles black; the eyes largely converging at the bottom, they being there not separated by much more than twice the width of the scape. Thorax black; the pronotum broadly, tegulæ and tubercles yellow. Mesonotum opaque; the central lobe raised; the scutellum yellow; the median segment reticulated; the pro- and meso-pleuræ longitudinally striolated, closely above, more widely below: the metapleuræ reticulated. The four front legs yellow; the tibiæ and femora infuscated beneath; the tarsi at the apex black. The wings reach to the middle of the abdomen. Petiole black, smooth and shining, the apical third dilated; its top with an elongated depression: the other segments testaceous beneath.

ANOMALON BRACHYPTERUM, *sp. nov.*

Nigrum, pedibus anterioribus, trochanteribusque posticis, pallidis; alis brevibus, hyalinis; abdomine testaceo, apice nigro. ♂. Long. 9 mm.

Hab. Trincomali, Ceylon (*Yerbury*).

Antennæ black, the scape yellow beneath; the flagellum covered with a microscopic down. Head black; closely punctured: the face densely covered with white hair: the mandibles testaceous, the palpi white. Thorax black, rough: in front sparsely, behind thickly, covered with white hair; the median segment with a gradual slope. Wings short, not reaching much beyond the apex of the petiole; the nervures black. The front four legs whitish yellow; the posterior black, the trochanters, knees, and spurs whitish-yellow (the front four legs are vermilion, but this is probably owing to discoloration with chemicals). Abdomen more than twice the length of the head: the petiole longer than the second segment, nodose at the apex; the base black; the apex brownish; the rest rufo-testaceous; the second segment testaceous, black above; the apical two segments black above.

Anomalon mussouriense, sp. nov.

Nigrum; flagello antennarum, pedibus abdomineque fulvis; alis fulvo-fumatis. ♀. Long. 17—18 mm.

Hab. Mussouri (*Rothney*).

Antennæ fulvous: the basal two joints entirely, and the third above, black. Head black; thickly covered with long fulvous hair, palest on the face; the lower three-fourths of the inner orbits, broadly in front, narrowed behind, the front and vertex coarsely, rugosely punctured: the face below the antennæ, the clypeus, labrum, and the mandibles, except at base and apex, fulvous-yellow: the palpi rufous: the face and clypeus coarsely punctured, depressed at the sides. Thorax strongly punctured; thickly covered with short fuscous hair; the scutellum yellow; a fulvous mark on the apex of the mesopleuræ and an oval one on the metapleura, rufo-fulvous: the suture on the apex of the mesopleura, yellow. The median segment coarsely reticulated; depressed in the middle: the apex in the centre with stout curved transverse keels. Legs rufous; the anterior paler, of a more yellowish tinge; the four posterior coxæ black, rufous at the base; the apex of the hinder femora and of the hinder tibiæ, black. A line on the top of the second and on the top of the fifth and sixth and the third to sixth abdominal segments, broadly at the sides on the lower part, black. The wings are uniformly fulvous smoky; the stigma and costa fulvous; the other nervures fuscous.

Campoplex buddha, sp. nov.

Niger, tibiis tarsisque anticis flavis; abdominis medio late rufo; alis hyalinis; nervis stigmateque nigris. ♀. Long. 14 mm.

Hab. Mussouri (*Rothney*).

Antennæ black, shining, sparsely covered with long white hair. Head closely and almost uniformly punctured, the face thickly covered with white hair; the hair on the top is equally thick and somewhat longer. Palpi testaceous, black at the base. Thorax closely punctured, thickly covered with white hair, short on the mesonotum, longer on the rest of the thorax; scutellum distinctly margined at the sides; the median segment longitudinally rugulose; the apex more coarsely transversely striolated. Propleuræ on the lower side irregularly obliquely striolated; the apex of the mesopleuræ shining, almost impunctate; the metapleuræ opaque, finely rugose. Legs covered with a white down: the anterior knees, tibiæ, and tarsi yellow; the middle knees testaceous; the calcaria white. Abdomen shining: the third and fourth and the lower half of the fifth segments rufous.

CAMPOPLEX SPECIOSUS, sp. nov.

Long. 12 mm.
Hab. Ceylon.

Is very near to C. buddha; but may be known from it by the base of the median segment having a clearly defined large triangular keel.

Head closely and uniformly punctured; thickly covered with glistening white hair; the mandibles yellow, the teeth piceous. Antennæ entirely black, longer than the body. Thorax black, closely punctured, thickly covered with short white hair; in the centre of the mesonotum is a longitudinal furrow: the median segment in the middle in the part below the triangular keel is irregularly striated; the apex with the striæ more apart, and it is more shining. The mesopleuræ have the punctures more distinctly separated than the others. The anterior coxæ and trochanters entirely, the apex of the middle coxæ and the basal joint of the middle trochanters and the underside of the four anterior femora, bright lemon yellow; the anterior tibiæ and tarsi entirely and the middle tibiæ in front whitish-yellow; the coxæ thickly covered with long glistening white hair; the calcaria white. Wings clear hyaline; the costa and nervures black. Abdomen black; the third to fifth segment red.

CAMPOPLEX SUMPTUOSUS, sp. nov.

Hab. Ceylon.

Is similarly coloured to the preceding two species, but is much smaller (7 mm.), and otherwise may be readily separated by the two keels at the base of the median segment being roundly curved.

Black; the apex of the second segment, the third and fourth and the base of the fifth segments red; wings clear hyaline. Head black, closely and uniformly punctured, thickly covered with short white hair, darkest and shortest on the vertex; mandibles and palpi black. Scape of antennæ sparsely covered with white hair; the flagellum with a close, black, microscopic down. Thorax closely punctured, the propleuræ strongly obliquely striolated at the bottom; the raised part on the mesopleuræ below the tegulæ finely transversely striated. Median segment broadly, but not deeply depressed in the middle, the basal keel roundly curved. Legs thickly covered with white microscopic down; the calcaria black. The areolet oblique, triangular at the top; the recurrent nervure received near the apical third of the areolet.

LIMNERIA CEYLONICA, sp. nov.

Nigra, abdominis apice late rufo; trochanteribus, tibiis, tarsisque anticis, rufis; alis hyalinis, stigmate nigro. ♂. Long. 7—8 mm.

Hab. Ceylon (Rothney).

Head very closely and rather strongly punctured all over; the face somewhat thickly covered with short white hair; the mandibles ferruginous, black at the base; the palpi testaceous, paler towards the apex. Antennæ entirely black, covered with a dark microscopic down. Thorax entirely black, alutaceous; thickly covered with white hair; the propleuræ shining, obliquely striated; strongly at the base, much finer at the apex: the mesopleuræ punctured: in the centre above to near the middle transversely striolated; the metapleuræ alutaceous; all thickly covered with short white hair; the median segment has a gradually rounded slope, and is thickly covered with white hair. Anterior coxæ black, white at the apex: the base of the trochanters and the anterior femora tibiæ and tarsi fulvous: the apex of the middle femora and base of tibiæ, rufous: all the spurs pale. Wings clear hyaline, slightly infuscated towards the apex: the areolet shortly appendiculated at the top: the nervures slightly curved, the lower side sharply angled in the middle. The basal segment of the abdomen entirely black; the second segment black, except the apex above and a mark on the side of the apex which are rufous like the rest of the abdomen.

LIMNERIA AGRAENSIS, sp. nov.

Nigra, pedibus rufis: apice tibiarum posticarum tarsisque nigris; alis hyalinis, stigmate testaceo; tegulis flavis. ♀. Long. 7—8 mm.; terebra 3 mm.

Hab. Agra (*Rothney*).

Head alutaceous, except on the vertex very thickly covered with white hair; the mandibles testaceous, thickly covered with golden hair; the teeth black, the part in front of them piceous; the palpi yellow. Thorax black; closely punctured, thickly covered with white hair; the middle of the mesopleuræ transversely striated, and with a smooth spot at the apex of the striated part; the basal area of the median segment larger, longer than broad; the keel straight, forming an acute angle in the centre; the apex of the apical area bulges into it as a triangle from the sides of which a keel goes round the edge of the segment; there is a short, stout, oblique keel outside the spiracles, beyond which it curves round to the apex of the segment, but is much thinner than the basal branch. Legs rufous; the base of the anterior pair yellow; the apex of the hind tibiæ and the tarsi black; the latter thickly covered with a white down, the spurs pale yellow. Wings clear hyaline; the stigma and nervures dark testaceous. Abdomen black; the second and third segments pale testaceous beneath; the petiole with an elongated area at the base of the thickened part; the apices of the second and third segments obscure rufous.

LIMNERIA MOROSA, *sp. nov.*

Nigra, palpis tegulisque albis; pedibus rufo-testaceis, abdominis segmentis testaceo-maculatis; alis hyalinis. ♀. Long. 5 mm.

Hab. Trincomali, Ceylon (*Yerbury*).

Antennæ entirely black, thickly covered with a pale microscopic pubescence. Head closely punctured, the face thickly covered with white pubescence; the mandibles and palpi white. Thorax shagreened, opaque, sparsely covered with minute pale hair; the three basal areæ on the median segment distinct; the others not clearly defined, the apex finely transversely striated. The four anterior legs pale testaceous; the femora with a more reddish hue; the coxæ broadly black at the base; the hinder entirely black; the apex of the hinder tibiæ and tarsi fuscous. Petiole black, shagreened, the base flat, very smooth and shining; the other segments black, broadly rufo-testaceous at the apex and at the sides; the ventral segments of a paler more yellowish testaceous colour. The stigma testaceous on the lowerside; the areolet distinctly petiolated.

PANISCUS CEYLONICUS, *sp. nov.*

Long. 19 mm.; terebra 4 mm.

Hab. Trincomali, Ceylon (*Yerbury*).

Comes near to *P. lineatus*, Bé. from Bengal, but that has the mesonotum marked with brownish lines; the alar nervures brownish at the base, reddish at the apex, while here they are uniformly black, and no mention is made of the dark antennæ.

Antennæ as long as the body; the scape testaceous; the flagellum black, dark brownish on the underside beyond the middle. Head clear yellow, the occiput in the middle of a more fulvous line; the teeth of the mandibles black, rufous at their base. Thorax bearing a microscopic white down; the median segment finely and closely but distinctly transversely striated. The hinder tarsi pale yellow. Abdomen infuscated towards the apex. The second transverse cubital nervure is interrupted on the lower side.

TRYPHONIDES.

EXOCHUS AITKINI, *sp. nov.*

Niger, pedibus stramineis; coxis tarsisque posticis nigris; alis hyalinis, stigmate fusco. ♂. Long. 7 mm.

Hab. Bengal (*E. H. Aitkin*).

Head shining, closely covered with short black hair; below the antennæ closely and somewhat strongly punctured; a semicircular furrow in front of the ocelli, the palpi, yellow; the mandibles before the apex piceous. Antennæ bearing a close fuscous pile; the flagellum, especially towards the base, brownish. Pro- and mesonotum sparsely covered with fuscous pubescence; the supramedian area on the median segment a little longer than broad; the keels at its base curved outwardly to shortly beyond the middle, when they become straight and oblique; the apical keel transverse. Pro-, meso-, and base of the meta-pleuræ shining and impunctate; the latter with a curved keel on the innerside of the spiracle, beyond which the segment is shagreened. The base of the petiole depressed; the depression margined, the margin continued shortly beyond it as blunt keels; the apex of the segment obscurely punctured.

The areolet is petiolated to near the bottom, where there is formed a minute cellule, not much wider than the transverse cubital nervure: its outer nervure is faint, and is interstitial with the recurrent.

BRACONIDÆ.

Bracon ceylonicus, sp. nov. (Pl. 3, f. 5).

Niger, pro- meso-thoraceque rufis; pedibus anticis testaceis; alis fere hyalinis. ♀. Long. fere 7 mm.; terebra fere 2 mm.

Hab. Ceylon (*Yerbury*).

Head black, shining, impunctate, the oral region (including clypeus) rufo-testaceous; the tips of the mandibles black; the palpi pallid testaceous. Antennæ longer than the body, entirely black. Thorax shining, impunctate, sparsely covered with white pubescence; the prosternum black: the metapleuræ and the median segment at the apex infuscated. Wings longer than the body: the lower side of the stigma fuscous. The middle legs, except at the base, infuscated; the hinder coxæ, femora, tibiæ, and tarsi, black; the trochanters fuscous. Abdomen black; the basal three ventral segments white, with a black spot in the centre of each. Petiole broadly depressed at the base and down the sides; the rest rugosely longitudinally punctured. The other dorsal segments coarsely rugosely punctured; the second with a raised somewhat triangular space in the centre at the base, from which a sharp keel runs to near the apex; at its side is an oblique furrow, with a sharp border on the inner side: the suturiform articulation longitudinally striated.

BRACON TRICARINATUS, sp. nov.

Niger, capite, prothorace, scutello, abdominis basi et apice pallide luteis; alis fere hyalinis. ♀. Long. 7 mm.; terebra 3 mm.

Hab. Ceylon (Yerbury).

Head testaceous-yellow, except at the orbits; shining, impunctate; the tips of the mandibles black and piceous; a deep, wide furrow leads down from the ocelli. The scape of the antennæ black: the flagellum broken off. Pro- and meso-notum smooth, shining, impunctate; in the centre of the latter is a large black mark reaching from the extreme base to near the middle, and two equally large lateral ones reaching from near the base to the apex; the scutellum luteous; the mesonotum at its sides and apex, black. Median segment entirely black; at the base in the centre is a depression which is finely longitudinally striated. Meso- and meta-sternum black: the metapleuræ and the mesopleuræ from the end of the oblique furrow, black. The front legs are entirely testaceous; the middle pair testaceous except the coxæ and trochanters; the posterior pair entirely black. The petiole testaceous; its raised centre black; the raised central part is narrowed gradually towards the apex, where it is a little less than the width of the lateral parts; at the apex it is stoutly keeled in the centre with the sides depressed; the base of the depression sharply keeled: the lateral depression on the inner side obliquely striated; its sides keeled down the centre. The second segment in the centre irregularly reticulated; the sides rugosely punctured; in the centre is a straight, stout keel, triangularly dilated at the base, this part being aciculated; the latter keels are equally stout, not dilated at the base and oblique; the third segment is longitudinally rugose, except at its sides at the apex, where it is smooth and shining; in the centre

c

is a keel; the other segments are only black down the centre and at the sides; the ventral segments black, the base testaceous, sharply produced in the middle.

BRACON ITEA, sp. nov.

Long. 4—5 mm.; terebra 2 mm.

Hab. Trincomali (*Yerbury*).

Head testaceous, shining, the face, except in the centre, aciculated; the palpi pale; antennæ black; the second joint obscure testaceous. Thorax testaceous, the median segment infuscated in the middle, where there is a shallow furrow. The legs pallid testaceous; the hinder femora and tibiæ infuscated, the former above and beneath. The raised central part of the petiole aciculated; the second and third segments rather strongly rugosely punctured; keeled down the centre; the dilated base of the keel on the second segment aciculated; the suturiform articulation finely longitudinally striolated; the third and fourth segments with a transverse furrow, oblique at the sides and longitudinally striolated; the second, third, and fourth segments broadly black in the middle; the black suffused with piceous on the third; the third with a distinct, the fourth with a less distinct longitudinal furrow; the ventral segments yellowish-testaceous.

BRACON AGRAENSIS, sp. nov. (Pl. 3. f. 6).

Flavus, vertice antennisque nigris; alis fuliginosis, basi late flavo. ♀. Long. 13 mm.; terebra 2 mm.

Hab. Agra (*Rothney*).

Antennæ as long as the body, black, almost glabrous. Head shining, thickly covered with long fulvous hair; the sides of the clypeus with an oblique, the base with a straight furrow; the apices of the mandibles black; behind the black extends to near the middle of the eyes. Thorax entirely yellow, smooth; a broad, curved furrow across the mesopleuræ. Legs entirely yellow. The raised part of the petiole strongly longitudinally striolated; the second segment inside the oblique furrows strongly longitudinally striolated; the base at the sides smooth; the furrows striolated; the suturiform articulation longitudinally striolated, broadened at the sides. The wings are yellow to near the base of the first cubital cellule; the first cubital cellule is hyaline above and beneath and at the base; and there is a clearer hyaline spot below the transverse cubital nervure. The stigma is broadly yellow at the base.

BRACON INGRATUS, *sp. nov.*

Long. fere 10 mm.

Hab. Agra (*Rothney*).

Head shining; the tips of the mandibles black; a broad furrow leads down from the ocelli. Thorax shining, impunctate; the curved furrow on the mesopleuræ wide; the median segment with a gradual slope, very smooth and shining. On the metapleuræ is a broad oblique furrow. Legs entirely luteous, the tibiæ thickly covered with pale hair. Wings bright yellow to near the stigma, which is luteous, black at base and apex; the first cubital cellule with a large somewhat triangular hyaline spot, and there is a smaller one below the first transverse cubital nervure. Petiole smooth; the apex in the centre with a few stout longitudinal keels; the lateral furrows wide. The other segments strongly rugosely punctured; the second segment with a stout keel in the centre reaching near to the apex; at the sides is a broad slightly curved depression, stoutly keeled on the innerside; the other segments have a stout transverse keel at the base, which become wider at the sides, and are crenulated.

In coloration it agrees exactly with *Bracon agraensis*; but may be at once separated from it by the strongly punctured abdomen with the longitudinal keel on the second segment.

Bracon Rothneyi, sp. nov.

Fulvus: alis fuscis, basi flavo; antennis nigris. Long. 6·5 mm.

Hab. Agra (*Rothney*).

Head entirely yellow, except the tips of the mandibles, which are black; the front and vertex shining, impunctate; below the antennæ it is obscurely rugose; furrowed down the centre; the clypeus shining, impunctate. Antennæ entirely black. Thorax above entirely smooth, shining, impunctate, very sparsely haired. Pleuræ smooth and shining; the metapleuræ with an oblique furrow. Legs entirely luteous, sparsely covered with white hair. Wings uniformly dark fuscous; the costa, except before the stigma, and the latter at the base, fulvous. Abdomen rugosely punctured; the raised part of the petiole with a double keel, open at the base, rounded at the apex. At the base of the second segment is a shining, smooth, raised area from which a stout keel proceeds to near the apex; at the side is a large oblique $\wedge$-shaped space, acutely margined on the innerside and obliquely striolated; there are indistinct depressions on the sides of the third and fourth segments.

Bracon Yerburyi, sp. nov.

Niger, orbitis oculorum, pro- et meso-thorace rufo-testaceis; tibiis, tarsis anterioribus, femoribusque anticis, testaceis; alis fere hyalinis. ♀. Long. 4 mm.; terebra fere 1 mm.

Hab. Ceylon, Trincomali (*Yerbury*).

Antennæ black: the flagellum covered with a very microscopic pile. Head obscure dark testaceous, darker on the face and on the vertex; the face with a distinct longitudinal keel. Thorax dark rufo-testaceous: the metathorax much darker; the mesopleura with an oblique deep wide furrow at the top. The hinder legs are entirely black and thickly covered with white microscopic pubescence; the middle tibiæ are obscure testaceous at the base; the rest of it and the tarsi black. The lower side of the stigma and the apical nervures are testaceous. The petiole obscure brown; the raised central part finely longitudinally striated; the second and third rather coarsely longitudinally striated; the second much more strongly than the third; and it has also its sides depressed and finely and irregularly striated; both have a smoother longitudinal line down the middle; the other segments are obscure brownish and aciculated.

SPINARIA NIGRICEPS, *sp. nov.* (Pl. 3, f. 7).

Nigra, thorace abdominisque basi rufis; pedibus anterioribus pallide flavis; alis fuscis. Long. 7 mm.

Hab. Ceylon (Yerbury).

Head black, the oral region and organs testaceous; a broad furrow leads down from the ocelli and there is a curved one over each antennæ. Antennæ longer than the body, black, the flagellum covered with a close microscopic pile. Thorax bare, shining; the sutures crenulated; a broad curved crenulated depression on the lower part of the mesosternum, which is black for the greater part; the median segment bears large shallow punctures. Legs covered with white hair; the four anterior entirely pallid yellow; the hinder black, the apices of the coxæ and the trochanters pale. Wings longer than the body; fusco-hyaline; the nervures and stigma blackish; the former paler towards the apex. Abdomen shining, base longitudinally striolated; the base with a large distinctly margined (rounded at the apex) space, which is smooth, except for a few scattered punctures; at the end of the metapleura over the apex of the hind coxa is a sharp tooth; the apical segment in the middle ends in two large sharp teeth, the part between them at their base being rounded; at their side is a shorter tooth about one-fourth of their length; the two proceeding segments end at the sides in large, sharp teeth. The basal segment is pallid fulvous, except for a black band in the middle; the second and third segments black, except the side of the second broadly and a triangular mark on the side of the third; the fourth segment is pallid rufo-fulvous except at the base and the sides: the terminal spines pale fulvous; the others deep black: the basal half of the ventral surface testaceous; the apical blackish.

The above described species comes nearest in form to *S. leucomelæna*, West., as also in its general coloration; but Westwood's species may be known from it by the black thorax.

APANTELES TAPROBANÆ, sp. nov.

Niger, pedibus abdominisque subtus rufo-testaceis; alis hyalinis. ♀. Long. 2 mm.

Hab. Trincomali, Ceylon (*Yerbury*).

Antennæ longer than the body, the scape testaceous, the flagellum obscure brownish beneath. Head black; the mandibles and palpi testaceous: the face finely punctured: its centre raised, the raised part becoming gradually wider towards the apex. Thorax above thickly covered with a pale pubescence; the median segment rugosely punctured: the propleuræ, the mesopleuræ in front of the depression, and the sternum finely punctured; the apex of the metapleuræ more closely and coarsely punctured; the legs rufo-testaceous except the base of the fore coxæ and the whole of the hinder coxæ, which are black: the latter coarsely punctured. The stigma fuscous; the nervures pale white, the basal two segments of the abdomen closely punctured: the sides of the second segment narrowly, the others broadly, and the ventral surface rufo-testaceous.

CHALCIDIDÆ.

CHALCIS BENGALENSIS, sp. nov. (Pl. 3, f. 9).

Nigra, pedibus anterioribus albis, basi late nigro; coxis trochanteribus femoribusque posticis, rufis; tibiis posticis albis, anticis nigris; tarsis posticis albis; alis hyalinis, nervis nigris; tegulis albis. ♀. Long. 2 mm.

Hab. Barrackpore (*Rothney*).

Antennæ 11-jointed, placed in the middle of the face; black; the scape bare, shining; the flagellum with a pale microscopic pile; the scape not reaching to the hinder ocelli; the antennal depression deep, sharply bordered; at the apex produced roundly in the middle; the vertex rough; the cheeks and clypeus covered thickly with long glistening white hair; the apex rounded at the top, smooth, and shining; the mandibles with the three teeth piceous. Thorax above coarsely punctured; covered sparsely with white hair; the sides and apex of the scutellum thickly covered with long silvery hair; the apex of the scutellum rounded; the median segment with an abruptly oblique slope; strongly reticulated. The lower part of the propleuræ coarsely, the upper part finely punctured; mesopleuræ coarsely punctured; the depression at its base wide, deep, strongly longitudinally striolated; the metapleuræ strongly irregularly reticulated, the reticulations much closer at the base. The four anterior coxæ and trochanters entirely black; the front femora black at the base; the middle black with the apex white; the hind coxæ (except at the base where they are black) trochanters and femora red; the tibiæ white, black in front; the tarsi white except at the apex; the femoral teeth, black, short, stout, closely pressed together at the apex, over a dozen in number.

CHALCIS ECCENTRICA, *sp. nov.*

Long. 5 mm.
Hab. Bombay (*Rothney*).

Very similar in coloration to *C. bengalensis*; but has

the scape of the antennæ reddish beneath; the thorax almost bare, the scutellum wanting the thick mass of white hair at the apex entirely; while the apex looked at from above is seen to be stoutly bidentate: instead of being uniformly rugosely punctured, the punctures are all widely separated, while at the base in the middle there is a large shining impunctate space.

Head rugosely punctured, sparsely covered with white hair; the clypeus shining, impunctate, glabrous, with two elongated punctures on either side of the middle: and there is a shining, impunctate spot above it. Base of mandibles finely longitudinally striated; the centre broadly rufous. Scape of antennæ rufous, darker at the apex; the flagellum stout, thickly covered with short white hair. Pro- and meso-notum with large deep punctures; the scutellum also strongly punctured: the centre at the base with a large smooth, impunctate space, surrounded by large shallow widely separated punctures; its apex ending in large teeth, rounded at the points; the median segment strongly reticulated. Propleuræ at top finely shagreened; its lower part and sides behind irregularly reticulated; the mesopleuræ shining; at the base on the lower side with some large deep punctures, this basal part being separated from the larger posterior by a distinct keel; the metapleuræ coarsely rugosely punctured, and in front thickly covered with long white hairs. Legs: the four anterior coxæ and trochanters, black; the four anterior femora broadly black at the base; the apex white: the tibiæ and tarsi white: the former broadly lined with black at the base: the hind coxæ, trochanters and femora red. The tibiæ and tarsi white, like the anterior, the tibiæ broadly black in the middle; the femora with 10 minute black teeth. Abdomen very smooth and shining; the penultimate segment aciculate, and bearing large deep round punctures.

HALTICELLA ERYTHROPUS, sp. nov.

Nigra, pedibus rufis, coxis, tibiis tarsisque posticis nigris; alis hyalinis. Long. 5 mm.

Hab. Agra (*Rothney*).

Head coarsely punctured; above sparsely covered with silvery hair; the face from the bottom of the eyes on either side of the antennal groove thickly covered with pale golden hair; the sides of the head stoutly margined. Antennæ long, slender, bare, the apex of the second joint rufous. Thorax strongly punctured, the punctures on the mesonotum much finer and closer at the base, and there is a smooth, impunctate spot on the sides. Median segment areolated, the base fringed with long pale golden hair: the basal central area elongated pyriform, transversely striolated. Propleuræ strongly punctured; the mesopleuræ hollowed, bare, stoutly longitudinally striolated; metapleuræ coarsely rugosely punctured, thickly covered with long fulvo-silvery hair. Fore coxæ with the edges on the outer side margined; the hind coxæ very smooth and shining, thickly covered with long silvery hair in front; the femora slightly, the tibiæ and tarsi thickly covered with silvery hair. Abdomen very smooth; the second segment at the top and apex laterally thickly and the other segments more sparsely covered with long silvery hair. Wings hyaline, a faint fuscous cloud under the costa; the nervures fuscous; tegulæ rufo-testaceous.

The parapsidal furrows are obsolete; the apex of the scutellum without teeth; the antennæ are 11-jointed, long, slender; the scape reaches to the ocelli; the hind coxæ have a large stout tooth at the apex. The fore tibiæ may be infuscated, and the hinder rufous behind; the median segment at the sides near the apex projects into a stout, large tooth.

This species agrees best with *Euchalcis* as defined by Kirby (*Journ. Linn. Soc.* (*Zool.*) xvii., 63).

TEMNATA, gen. nov.

Antennæ 12-jointed, situated immediately over the mouth. Face broadly, but not deeply, excavated. Mesonotum without parapsidal furrows. Scutellum at the apex narrowed, and projecting at the sides into two oblique triangular teeth. At the base of the metapleuræ near the hind wings are two stout keels almost united on the outer border and forming a somewhat horseshoe-shaped area: on the side beyond this are two stout spines widely separated. Ovipositor short.

Comes nearest to Kirby's genus *Megalocolus* (*Journ. Linn. Soc.* (*Zool.*) xvii., 61), which differs from it in having the antennæ inserted in the middle of the face; the hind coxæ have a leaf-like projection on the upperside, and the ovipositor is as long as the abdomen itself.

TEMNATA MACULIPENNIS, sp. nov. (Pl. 3, f. 10).

Nigra, argenteo-pilosa; alis fumatis, albo-fasciatis. ♀.
Long. 6 mm.

Hab. Agra (*Rothney*).

Head strongly punctured, very sparsely covered with a microscopic pile, which gives it a greyish appearance. From the middle of the lower side of the cheek a distinct keel runs to the eyes. Pro- and meso-notum closely punctured: the scutellum with the punctuation equally strong, but closer: the apical teeth are not twice longer than wide and rounded at the apex. Median segment with keels all over from the base to the apex: the two central straight, the others more oblique; the sides at the apex thickly covered with long silvery hair. Pro- and metapleura coarsely and uniformly punctured; the mesopleuræ with longitudinal keels rather widely separated: the upper side at the apex rugosely punctured. Legs black, sparsely covered with a silvery pile. Abdomen shining; the apex opaque, shagreened; the fore wings to the base of the stigma hyaline; there is then a narrow fuscous stripe, followed by a hyaline one extending a little beyond the cubitus: the rest of the wing smoky, lighter at the apex.

PROCTOTRUPIDÆ.

EPYRIS AMATORIUS, *sp. nov.* (Pl. 3, f. 8).

Long. 7 mm. ♂.

Hab. Barrackpore (*Rothney*).

Head strongly punctured, more widely separated behind the ocelli, the clypeus stoutly keeled down the middle, and a curved keel on either side of this united to the central at the base; the mandibles with large punctures; the four basal teeth brownish. Antennæ entirely black: the scape sparsely; the flagellum more closely covered with pale fuscous hair, nearly as long as the thickness of the joints. Pro- and meso-notum thickly covered with long fuscous hairs; the base of the pronotum closely transversely striolated; the rest of it coarsely irregularly rugosely punctured, except at the apex, the smooth apical part being separated by a distinct keel from the rest; the mesonotum with scattered punctures: the parapsidal furrows reaching not quite to the apex: the scutellum almost impunctate. The median segment transversely, more widely at the base, where there is in the centre a somewhat triangular area; the apex is more strongly and closely transversely striolated. Propleuræ shining, smooth; the mesopleuræ covered with large, distinctly separated punctures, except a smooth, elongated, slightly raised space under the wings; the metapleuræ punctured at the top and round the apex; the top at the base with two longitudinal keels, between which are two perpendicular ones. Legs black, the joints testaceous; the femora and tibiæ sparsely, the tarsi more thickly covered with shorter white hair. Wings hyaline, with a very faint fulvous tinge: the stigma black; the nervures testaceous. Abdomen shining; the apex sparsely covered with long pale hair; at the base is a distinctly bordered longitudinal furrow.

Except the radius, the alar nervures are obsolete.

SCOLIIDÆ.

Tiphia tarsata, sp. nov.

Nigra, tibiis tarsisque anticis rufis; alis fusco-hyalinis. ♂. Long. 9 mm.

Hab. Mussouri (*Rothney*).

Head shining, strongly punctured; the punctures widely separated on the vertex: the front and vertex covered with longish pale hair; the clypeus and lower part of the cheeks thickly covered with long white hair; the base and the apex of the mandibles broadly in the centre, ferruginous. The scape of the antennæ strongly punctured beneath and sparsely covered with long white hair; the flagellum obscure brownish beneath. Pronotum punctured, except at the apex; and rather thickly covered with long pale hairs; the mesonotum with a broad fringe of large punctures round the sides; the scutellum with large punctures all over, which are much closer towards the apex. Median segment alutaceous: with three complete keels down the centre; the apex at the sides shining, smooth. Propleuræ obscurely punctured round the edges: the rest finely obliquely striated: the mesopleuræ strongly punctured; the metapleuræ obliquely, somewhat irregularly striated, the striæ widely separated. Legs thickly covered with white hairs; the fore knees, femora, and tarsi rufous; the tarsi pale; the apex of the middle tibiæ and the middle tarsi testaceous, as are also the apices of the basal two joints and the third joint of the posterior tarsi. Abdomen shining, thickly covered with long white hair, especially towards the apex, where it has a fulvous hue.

Tiphia Magrettii, sp. nov.

Nigra, nitida, femoribus posticis rufis : alis fusco-hyalinis.
Long. 10 mm.

Hab. Mussouri (*Rothney*).

Black, shining, sparsely covered with longish glistening white hair. Antennæ entirely black; the scape shining, bearing a few large punctures and long white hairs; the flagellum opaque, covered with a dull microscopic pile. Head covered with long glistening white hair and bearing large moderately deep punctures; the mandibles shining, grooved, and broadly red towards the middle. Pro- and meso-notum shining, bearing long white hairs; and widely separated punctures; the scutellum irregularly punctured round the sides and apex; those on the latter being the larger; the apex of the post-scutellum with scattered punctures. Median segment shagreened; the base almost glabrous, with two complete keels in the centre, and having between them one which is only three-fourths of their length; the apex slightly hollowed towards the centre, which has a straight keel; sparsely covered with long white hair. Propleuræ smooth, shining above; the lower part obscurely transversely striated ; the mesopleuræ projecting at the base; almost straight, smooth and impunctate; the sides rather strongly punctured, but with the punctures all distinctly separated; sparsely covered with long white hairs; the metapleuræ strongly obliquely punctured. Legs thickly covered with stiff white hairs; the short thick spines on the hind tibiæ and the calcaria pallid testaceous. From the stigma the wings have a decided smoky tinge; and are traversed by four white lines (two above and two below the cubital nervure) like nervures. Abdomen black, shining, and covered, especially towards the apex, with long white hairs; the sides of the basal segments sparsely; the apical more closely and thickly covered with long white hairs; the last segment more or less piceous; the ventral segments shining, sparsely covered with long white hairs.

This can hardly be *Tiphia rufofemorata* Sm., for the head with "numerous fine punctures," "the apical half of the mandibles ferruginous," the scutellum "strongly punctured" cannot apply to our species; nor is there any mention of the metasternum being striolated. It is very like the well-known European species *Tiphia femorata*; but differs in having the hinder tibiæ black; the apex of the clypeus more sharply projecting and more deeply incised, and the second abdominal segment not depressed and crenulated at the base.

Tiphia femorata is recorded by Magretti from Burmah. (*Ann. Mus. Civ. Genova*, XII. p. 52.)

Tiphia cassiope, sp. nov.

♀. Long. 7 mm.

Hab. Mussouri (*Rothney*).

Resembles *T. Magrettii* in coloration, but is smaller and the clypeus is not distinctly projecting and incised in the middle as it is in *T. Magrettii*.

Head shining, sparsely punctured, and bearing some long white hairs behind; the clypeus punctured, transverse, the apex smooth; the apical three-fourths of the mandibles rufous, the extreme apex black. Antennæ black; the apex of the scape piceous; bearing a few large punctures and some longish pale golden hairs. Pro- and meso-notum shining; the former with the basal three-fourths punctured, the punctures being closer together at the base; the sides in the middle and the apex, impunctate; the mesonotum with moderately large punctures in the middle, the sides with a few widely separated punctures; the scutellum punctured at the apex and sides. Median segment shining, slightly shagreened at the base; the three longitudinal keels complete. The lower half of the propleuræ finely longitudinally striated; the mesopleuræ with scattered punctures and sparsely covered with long white hairs; the metapleuræ finely and closely longitudinally striated. Legs thickly covered with long white hairs, the four hinder trochanters and femora bright rufous; the fore femora and tibiæ underneath more or less dull rufous. Wings hyaline, slightly suffused with fuscous. Abdomen shining, the apex of the first segment with a transverse row of punctures; the apical half of the last segment shining, dull piceous.

The three species here described with red on the legs may be separated as follows:—

1 (4) Femora red, wings smoky.
2 (3) Clypeus incised, projecting. *Magrettii*
3 (2)   ,,   transverse, not projecting. *cassiope*
4 (1) Femora black; anterior tibiæ and tarsi
         rufous. *tarsata*

TIPHIA CLYPEALIS, sp. nov.

Nigra, clypeo, tibiis anticis, tarsisque anterioribus, rufis; alis hyalinis, nervis fuscis. ♀. Long. 7 mm.

Hab. Mussouri (*Rothney*).

Head shining, punctured; covered with long silvery hairs, which are densest below the antennæ; the clypeus rufous; punctured; the apex smooth, impunctate, and slightly curved; mandibles broadly rufous in the middle. Antennæ ferruginous beneath towards the apex, the last joint entirely so. Pronotum shining, punctured; the mesonotum with the punctures more widely separated; the sides being free from them, and being there too more widely separated than they are on the sides; the scutellum with large, widely separated punctures all over. Median segment coarsely in the middle, the sides much more finely aciculated. Propleuræ finely obliquely striated throughout. Legs thickly covered with white hairs; the anterior knees, tibiæ, and tarsi, the base and apex of the middle tibiæ and the apices of the hinder tarsal joints, rufous. Radial cellule closed; the second recurrent nervure received in the apical third of the cellule. Basal segment of the abdomen except a belt at the apex, with only a few scattered indistinct punctures; the apices of the others closely punctured; and sparsely covered with long white hairs; the apical ventral segment strongly aciculated, rufous at the apex; there is a distinct curved keel on either side of the penultimate segment.

✓ Tiphia fuscinervis, sp. nov.

Nigra, tarsis anticis rufis; abdominis apice longe fulvo-hirto: alis hyalinis, nervis stigmateque fuscis. ♀. Long. fere 8 mm.

Hab. Mussouri (*Rothney*).

Head densely covered with long fuscous hairs; shining, strongly punctured, the mandibles broadly ferruginous before the apex; the palpi dark testaceous. Antennæ obscure brownish towards the apex, covered with a pale microscopic pile, the scape shining, coarsely punctured on the inner side. Pronotum closely punctured; the mesonotum with the punctures larger and more widely separated; the scutellum with a wide belt of punctures at the apex, a narrower one at the sides and base, and a somewhat broader one down the middle; the post-scutellum finely rugose. Median segment coarsely alutaceous; the keels straight, a little converging towards the apex; an interrupted keel down the middle at the base; the apex with an oblique slope. The lower part of the propleuræ obliquely striolated; the upper part obliquely aciculated; the mesopleuræ strongly punctured. Legs thickly covered with longish white hairs; the calcaria pale luteous. Wings hyaline; the stigma dark piceous; the nervures pale testaceous. The basal segment of the abdomen very smooth and shining, sparsely punctured in the middle; the third and following segments punctured, thickly covered with long pale fulvous hairs, which are more silvery towards the apex.

TIPHIA INCISA, *sp. nov.*

Nigra, longe argenteo-pilosa; apice clypei incisa: alis hyalinis, nervis fuscis. ♂. Long. 9—10 mm.

Hab. Mussouri (*Rothney*).

Head black, thickly covered with cinereous pubescence; rather strongly punctured; the clypeus closely punctured; the apex smooth, roundly incised. Antennæ thick, the scape with a few large punctures and with longish white hairs; the flagellum closely covered with a pale microscopic down. Pronotum closely punctured, the extreme apex only impunctate; the mesonotum strongly punctured but not closely; the scutellum more closely punctured all over, this being also the case with the post scutellum; the median segment short, finely rugose, opaque, the base sparsely, the apex much more densely covered with long white hairs; at the base are two straight keels, with an indistinct one in the centre, the two forming an area nearly as broad as long; the apex has an oblique slope and has an indistinct keel down the centre. Propleuræ aciculated, obscurely striated at the bottom; mesopleuræ punctured; the metapleuræ with about eight semi-oblique keels at the top. Wings hyaline, the nervures and stigma black; the radial cellule closed at the apex. Legs entirely black except the calcaria, which are pale fulvous; the tarsi with a fulvous pubescence beneath. Abdomen covered with longish white hairs; the basal segment with widely separated punctures all over, its apex depressed; the apical more closely and strongly punctured than the middle segments.

Tiphia implicata, sp. nov.

Long. 9 mm.

Hab. Mussouri (*Rothney*).

Head densely covered all over with long white hair: opaque, the clypeus largely produced and projecting ; the sides oblique, the apex transverse ; the mandibles entirely black, covered with long white hairs ; the base punctured. Pro- and meso-notum closely punctured all over; the latter more strongly than the former, and more sparsely towards the middle ; the scutellum punctured all over like the mesonotum ; the post-scutellum closely finely rugosely punctured. The median segment finely rugosely punctured, towards the apical keel irregularly striolated : the two outer keels curving inwardly ; the central straight, not reaching quite to the apex. The apex of the segment sharply oblique. Propleuræ strongly transversely striolated, except at the extreme apex, which is shining and impunctate ; the mesopleura alutaceous ; the metapleura depressed at the base, closely longitudinally striolated. Legs black, thickly covered with long pale hairs, the hairs on the underside of the tarsi pale golden : the calcaria and the tibial and tarsal spines pale fulvous. The wings hyaline, infuscated towards the apex ; the radial cellule is not appendiculated ; the second recurrent nervure is received shortly beyond the middle of the cellule. Abdomen shining, impunctate at the base, more opaque and thickly haired towards the apex ; the basal segment above with a long shallow depression ; the ventral segments sparsely covered with long pale hairs.

Tiphia erythrocera, sp. nov.

Nigra, mandibulis, tibiis, tarsis anticis, flagelloque antennarum rufis; alis hyalinis. ♀. Long. 8 mm.

Hab. Mussouri (*Rothney*).

Antennæ rufous, covered with a pale microscopic pile ;

the base of the scape black; covered with long golden hairs on the underside. Head shining, sparsely haired: covered with large distinctly separated punctures: the mandibles ferruginous; sparsely covered on the lower side with long golden hairs; the teeth are black. Pro- and meso-notum with scattered punctures except at the apices; the scutellum with a few punctures at the apex. Median segment alutaceous; the three keels complete. Propleuræ alutaceous, smooth and shining above: the mesopleuræ punctured, alutaceous at the top: the metapleuræ striolated throughout, much more finely at the base. Legs thickly covered with white hairs: the tibial spines pale; the four anterior tarsi; the front tibiæ behind and the middle tibiæ entirely ferruginous; the hinder tarsi ferruginous; the calcaria and tarsal spines pale fulvous. Wings hyaline, suffused with fuscous; the nervures fuscous; the stigma black; the second recurrent nervure received in the apical third of the cellule. Abdomen shining, sparsely covered with long white hairs; the transverse depression on the apex of the petiole closely and coarsely punctured at the sides, more widely and sparsely at the middle; the puncturing on the dorsal segments becomes closer and coarser towards the apical; the last shining, impunctate, piceous broadly at the apex.

METHOCA.

Smith described two Indian species of *Methoca*, under the same name—*orientalis*—(Cat. Hym. III., 66) from Northern India and another, renamed *Smithii* by Magretti (*Ann. Mus. Civ. Genova, xxxii., p. 259*), taken by Mr. Rothney at Barrackpore (*Trans. Ent. Soc., 1875, p. 35*). Both were described from males: and represent, so far as can be judged from the descriptions, different species. The undernoted female is, I should say, quite distinct from either.

Methoca bicolor, *sp. nov.* (Pl. 4, f. 12).

Nigra, nitida, thorace basique abdominis rufis. ♀. Long. 5 mm.

Hab. Barrackpore (*Rothney*).

Antennæ stout, the four basal joints of the antennæ rufous, sparsely covered with white hairs, becoming slightly and very gradually thickened towards the apex. Head shining, impunctate. Thorax shining, impunctate, except at the base of the scutellar region where it is transversely striated; the mesonotum at the sides of the scutellum is also somewhat obliquely strongly striated. The basal segment of the abdomen is rufous, except at the extreme apex; the other segments shining, impunctate; the apical segment obscure rufous. Legs black, the tarsi obscure testaceous, the femora sparsely haired; the tibiæ covered with stiff hairs.

Methoca rugosa, *sp. nov.* (Pl. 4, f. 11).

Nigra, basi flagello antennarum late, femoribus tarsisque anticis rufis; alis violaceis, basi hyalinis. ♂. Long. 15 mm.

Hab. Ceylon.

Antennæ stout, almost bare, the basal three joints and the base of the fourth rufous. Head black; the mandibles broadly rufous in the centre; the front strongly punctured, almost reticulated, thickly covered with fuscous hairs; the vertex more shining, less pilose; the punctures shallower and more widely separated, especially at the side of the ocelli. Thorax black, the pronotum, except a smooth, impunctate band at the apex, coarsely transversely striolated; the mesonotum much more strongly and irregularly transversely striolated; scutellum strongly irregularly reticulated; the sides towards the apex impunctate; in the centre of the metanotum is a pear-shaped area, with four stout transverse keels, the two central being the longest; at the side of this are stout semi-longitudinal keels; the apex at the top is stoutly margined; at the top is a triangular area, the sides with stout oblique keels, meeting in the centre; the pleura coarsely irregularly reticulated, the sternum irregularly transversely striolated; its side stoutly keeled with a sharp margin at the edge, the pleura at the side of this being hollowed; the sternum widely hollowed, the hollow becoming gradually wider towards the apex. Abdomen shining, impunctate; the base stoutly longitudinally striated; the basal ventral segment strongly reticulated; keeled down the middle to near the apex: the other segments with punctures at the apex, these being fewer on the middle and more numerous on the apical segments.

MUTILLIDÆ.

Since my paper on the Indian Mutillidæ (*Manchester Memoirs, V., 1892*) was published, some additional species have come into my possession from Mr. Rothney and from Col. Yerbury. The collection from the last-named gentleman is of especial value, as it enables us to unite the sexes of a few species.

The discovery by Mr. Rothney of an apterous ♂ *Mutilla*, although not unique, is of interest. It is remarkable that the four known apterous species of ♂ *Mutilla* have the thorax emarginate, as it often is with ♀ *Mutillæ*, while it never is so in the winged males.

a. Males. Wingless, thorax incised.

MUTILLA ŒDIPUS, *sp. nov.* (Pl. 4. f. 13. ♂.)

Ferruginea, aptera, abdomine nigro, albo maculato; pedibus nigris; thorace late inciso. ♂. Long. fere 9 mm.

Hab. Barrackpore (*Rothney*).

Head large, wider than the thorax, the part behind the eyes more than twice their length; coarsely punctured, closely covered with white pubescence; black, ferruginous from shortly above the antennal tubercles to the top of the eyes; the black above the outer side of the eyes being oblique. Antennæ entirely black; the scape thickly covered with stiff white pubescence, the black part of the head densely covered with glistening white pubescence; the vertex and occiput with the pubescence longer, darker, and more erect. Thorax not twice the length of the head; gradually narrowed to the metathorax, which bulges out, so that it is as wide as the prothorax. Above, the thorax is coarsely rugose and covered with long fuscous hairs; the apex of the median segment is oblique and has a sharp spine in the centre. The pleuræ are shining, impunctate; covered with white pubescence; black; the upper part of the pro-, the upper third of the meso-, and the meta-pleuræ above the oblique furrow, rufous. The sides of the median segment with three large and one short spine. The legs black; covered densely with white hairs; the tibiæ almost spineless. The abdomen longer than the head and thorax united, velvety black, covered with long black hairs; a square spot on the centre of the first segment; three large oval ones at the apex of the second; a small one in the centre of the third, a larger one in the centre of the fourth, both narrowed and rounded at the base; and the greater part of the fifth, white. Ventral segments black; the second strongly punctured, sparsely covered with white hairs; the others are fringed with long pale hair.

The genital armature is normal.

b. Winged, thorax not incised.

Fore wings with only one recurrent nervure; three transverse cubital nervures; stigma elongate; apex of abdomen bispinose, middle tibiæ with two spines. Petiole serrate beneath, elongate, nodose at apex. Eyes very large, oval, entire; ocelli large.

The precise generic position of this species must stand over for further study in connection with its unknown ♀. It is very closely related to the *Photopsis* section of the American genus *Sphærophthalma*. In general form and appearance it is very like *Mutilla obliterata* Sm., which is, however, abundantly distinct otherwise. It differs from *Mutilla* proper in the eyes being entire in the ♂.

MUTILLA APICIPENNIS, *sp. nov.*

Thorace capiteque ferrugineis; abdomine nigro, basi ferrugineo; pedibus pallide testaceis; alis hyalinis, apice fumatis. ♂. Long. 10 mm.

Hab. Trincomali (Yerbury).

Head as wide as the thorax, shining, glabrous; the mandibles with long fulvous hairs; their teeth deep black; a slight depression on the front above the antennæ; the eyes and ocelli large; the head behind the eyes not half their length. Antennæ thick, uniformly fulvous, covered with a close white pubescence. Thorax uniformly fulvous, shining; the mesonotum obscurely punctured; the scutellum rugosely punctured; the median segment with a gradually rounded slope; reticulated uniformly, sparsely covered with long white hairs. Pro- and meso-pleuræ rugosely punctured; the edges of the former crenulated; the metapleuræ reticulated. Legs pale testaceous, covered with long pale hairs; the hinder femora broadly infuscated towards the apex. Petiole elongated, gradually dilated, and strongly punctured, especially towards the apex; where there is a black band; beneath it is hollow, shining, the edges rough, the other segments shining, their apices obscure testaceous; covered with long pale hairs. The wings, which do not reach to the apex of the abdomen, are milk-white. The nervures pale testaceous. The apex from the third transverse cubital nervure smoky; the two basal transverse cubital nervures curved; the third sharply angled: the first recurrent nervure received in the basal third of the cellule; the second completely obliterated.

The ocelli are larger than usual; the second abdominal segment unarmed beneath. On the mesonotum the two parapsidal furrows are complete; the last dorsal abdominal segment rufous and punctured at the apex.

In appearance this species resembles the American genus *Photopsis*. It is apparently closely related to *M. pedunculata* from Arabia and Egypt.

A. Descriptions of species known in both sexes.

The following species belongs to a group the species of which, being so similarly coloured, are very difficult to identify; and I should not have ventured to describe it if I had not got both sexes.

MUTILLA ACIDALIA, sp. nov.

♀. Black, thorax above ferruginous. Head as wide as the thorax, coarsely punctured: the head behind the eyes developed a little less than the length of these latter; covered with a short, sparse, white pubescence. Antennæ stout, covered with a white down; the basal joint reddish at the apex. Thorax above, coarsely punctured, sparsely covered with fuscous hairs; the median segment with an abrupt slope, coarsely punctured, covered with long white hairs. Abdomen not much longer than the head and thorax united; black; sparsely covered with long fuscous hairs; the hypopygium rather strongly longitudinally striated: on the second segment are two oval, on the third and fourth segments two square marks of silvery pubescence; the basal ventral segment is ferruginous; the others obscure testaceous at their apices: and marked with long white hairs. Legs black, the tibiæ and tarsi with white hairs; the tibial spines stout, fuscous; the calcaria pale.

♂. Head and thorax black, the former in front and the pronotum thickly covered with long white silvery hairs; abdomen ferruginous, except the apical segment, which is black above and beneath; wings fuscous, paler at the base. Antennæ elongate, slender, tapering towards the apex; the scape grooved laterally, sparsely covered with long white hair; the flagellum covered with a sparse down; the third joint nearly twice the length of the fourth. Head not much narrower than the thorax; behind the eyes it is a little longer than their width; the front and vertex strongly punctured. Prothorax strongly punctured, the pronotum thickly covered with grey pubescence; the mesonotum strongly punctured, the punctures deep and clearly separated; down the sides run two deep furrows; the median segment with a somewhat abrupt rounded slope: reticulated; in the centre is an elongated area reaching from the base to shortly beyond the middle, the base being dilated. Propleuræ obliquely striolated, smooth behind; the mesopleuræ coriaceous, projecting in the middle, where they are thickly covered with long white hairs. Legs covered with longish white hairs; the calcaria pale. Radial cellule elongate; the basal abscissa of the radius sharply oblique, the apical more rounded: the first transverse cubital nervure oblique; the second broadly and roundly curved: the third sharply angled above the middle; both the recurrent nervures are received shortly beyond the middle. Keel on basal ventral segment stout, black.

MUTILLA OPULENTA, Smith.

The ♀ of this species is probably *M. soror*, Sauss. (*Ann. Soc. Ent. Fr.*, vii., *1867, 354, t. 8, f. 3*.) As Smith's name is the older one, it will have to be adopted should *soror* prove to be a variety.

Col. Yerbury has taken the sexes together. The ♀ however has the thorax black, while in the typical *M. soror* it is reddish; but in other respects the two agree, except that the hinder femora have more black. It may, however, be as well to give a description of the ♀ *M. opulenta* in case *M. soror* may be different.

Head as wide as the thorax; red, coarsely and rugosely punctured, shining, bare, behind the eyes the vertex almost as long as the width of the eyes. Scape of the antennæ rufous, darker at the apex and beneath, and bearing large punctures; the flagellum black, thickly covered with fuscous pile. Thorax black, twice the length of the head; the sides almost parallel, not dilated towards the apex, very coarsely rugosely punctured; the apex of the median segment with a very slight oblique slope; above coarsely punctured, the rest finely and uniformly rugose; the lower part covered with long golden hairs. The pleuræ smooth; the base of the pronotum and the metapleuræ coarsely punctured; the lower portion of the metapleuræ thickly covered with pale golden pubescence. Petiole black, the apex (probably the whole in fresh examples) fringed with golden hairs; on the base of the second segment are two large oval golden marks; its apex has a golden band, broadly narrowed in the centre; the third segment is entirely golden; the pygidium coriaceous, fringed at the sides with long golden hairs. The second ventral segment has large, somewhat shallow, clearly separated punctures; all the segments fringed with long golden hairs. The front four legs red, the knees black, sparsely covered with golden hairs, especially the tarsi; the apical three-fourths of the hind femora are black; the tibiæ and tarsi thickly covered with golden hair.

The form of the spots on the second abdominal segment varies. In one example on the inner side at the apex they are rounded; in another they are there truncated as figured by Saussure in his *M. humbertiana;* in another they are more as he figures them in *M. soror*, but the band on the fourth segment is as in *M. humbertiana;* in *M. soror* it is figured as straight at the base. *M. insularis* Cam., may be known from it by the thorax being red; by the metapleuræ not being coarsely punctured throughout, only at the extreme apex, the band on the third segment complete, not incised at the base.

MUTILLA PERELEGANS, sp. nov.

This is the supposed variety of *M. pulchrina* figured by me (*Manchester Memoirs*, V., *1892*, pl. 1, f. 6), but which I now regard as quite distinct; and, thanks to Col. Yerbury, I am enabled to describe the male, as well as the female, in detail.

♀. Head very slightly narrower than the thorax; stoutly keeled on the sides behind, ferruginous, coarsely rugosely punctured; covered sparsely with longish black hairs; almost transverse behind, where it is developed the length of the eyes. Mandibles black. Scape of the antennæ deep black, shining, glabrous; the flagellum thick; the third joint twice the length of the fourth; brownish beneath. Thorax a little narrowed from the middle to the apex; above coarsely rugosely punctured, the punctures elongated; sparsely covered with long black hairs, but very thickly on the pronotum, while the median segment is thickly covered with long pale golden hairs, and has a somewhat oblique slope. Pro- and meso-pleura shining, impunctate; except a broad punctured projection down the mesopleuræ, the projection itself being covered with long pale golden hairs; and, above, it forms a projecting tooth, behind which is another slightly larger and rounder one. Legs black; the femora slightly, the tibiæ and tarsi thickly covered with long golden hairs. Abdomen longer than head and thorax together; black; an orange-coloured mark of hairs, broader than long and with the sides rounded, in the centre of the second segment; the sides of the second segment above broadly fringed with pale golden hairs: the third segment entirely covered with golden hairs; the rest of the abdomen with black stiff hairs: the third and fourth ventral segments covered with golden hairs; the fifth and sixth slightly fringed with golden hairs; the apical segment with a dense tuft of golden hairs at the end.

Length, 12 mm.

♂ much larger (17 mm.) has the head and thorax red: the abdomen with the apex of the third and the fourth, and the fifth segments entirely covered with golden pubescence. Head distinctly narrower than the thorax, coarsely rugosely punctured; sparsely covered with long black hair; below the antennæ the hairs are longer and fulvous; behind rounded at the sides, about one half the length of the eyes; the apical half of the mandibles black. Antennæ short, thick; but tapering very considerably towards the apex: the basal two joints red: the rest black and almost bare: the scape with a few long hairs: and strongly punctured above. Pro- and meso-notum coarsely rugosely punctured; thickly covered with long black hairs; scutellum flat, the sides and apex projecting: covered with long black hairs, except at the apex, where they are longer and pale fulvous in colour; this being also the case with the post-scutellum. Median segment strongly reticulated; the apex roundly emarginate; the sides projecting into stout teeth. Pro- and meso-pleuræ coarsely rugosely punctured except the lower part of the former. The fore legs are reddish like the thorax: the four hinder legs are entirely black, except the coxæ at the base; they have the femora slightly, the tibiæ and tarsi densely covered with long ' pale golden hairs. Wings fusco-violaceous: the base much lighter, almost hyaline; the basal and apical abscissæ of the radius are oblique and, at the base of the latter, it projects a little; the first and second transverse cubital nervures are curved ; the third is obliterated entirely, while the cubital nervure itself terminates at the second transverse cubital. The abdomen has the basal two segments strongly punctured ; golden band on the apex of the second segment is interrupted in the middle ; the basal ventral segment is more or less rufous, and projects at the apex into a sharp, triangular plate : at the base in the middle it is semi-circularly incised.

B. Species described from males only.

a. Fore wings with only two transverse cubital nervures.

MUTILLA PERVERSA, *sp. nov.*

Nigra, thorace rufo, sterno nigro ; alis subfumatis, nervis fuscis. ♂. Long. 5 mm.

Hab. Barrackpore (*Rothney*).

Head black, shining, sparsely covered with longish pale hairs. Vertex behind the eyes equal to their length, and not much narrowed ; the mandibles dark piceous, the teeth black ; the palpi fuscous. The antennal scape not furrowed beneath, sparsely haired ; the flagellum stout, covered with a microscopic down. Thorax above entirely obscure ferruginous, punctured, but not strongly ; sparsely covered with long white hairs ; the median segment with a gradually rounded slope. Propleuræ almost black ; the mesopleuræ obscurely punctured in front ; the metapleuræ impunctate at the base ; the apex strongly reticulated. Legs covered with long soft hairs ; the calcaria white. Alar nervures fuscous ; the first abscissa of the radius oblique ; the apical small, almost straight ; the first transverse cubital nervure straight, oblique ; the second curved and largely bullated at the bottom ; the first recurrent nervure is received shortly beyond the middle of the cellule. Abdomen shining, almost impunctate covered, especially towards the apex, with long soft white hairs. The basal abdominal segment without a keel sharply separated from the second, which is gradually raised to the obliquely depressed apex, thus leaving a sharp depression between the two.

b. Fore wings normal, with three transverse cubital nervures.

MUTILLA INDEFENSA, *sp. nov.*

Nigra, vertice fulvo-hirta; collare late abdomineque ferrugineis; abdominis basi apiceque nigris; alis fuliginosis. ♂. Long. 17 mm.

Hab. Bombay District (*Wroughton*).

Head a little narrower than the thorax; the vertex and front densely covered with fulvous pubescence, intermixed with long fuscous hair; the clypeus sharply keeled in the middle; the mandibles entirely black, fringed with long golden hairs. Antennæ entirely black, the scape widely grooved; sparsely covered with long fuscous hairs. The head behind the eyes is rapidly narrowed, and is not half the length of the eyes. The pronotum is broadly covered with thick orange pubescence; the pleuræ at the base coarsely punctured; the rest is longitudinally striolated. Mesonotum coarsely punctured especially towards the apex: there are two moderately wide longitudinal furrows. Scutellum pyramidal; coarsely rugosely punctured; and, like the mesonotum, thickly covered with long black hairs. Median segment coarsely reticulated; its base thickly covered with golden hairs; the centre with an elongated area reaching to the edge of the slope, which is oblique. Mesopleuræ coarsely punctured, covered with silvery pubescence; the apex impunctate; the base of the metapleuræ impunctate; the apex coarsely reticulated. Legs black; the calcaria white; the femora, tibiæ, and tarsi thickly covered with long white hairs. Wings fusco-violaceous, paler at the base; the basal abscissa of the radius oblique and twice the length of the apical; the second cubital cellule elongate, its lower side twice the length of the third ; the first recurrent nervure is received shortly beyond the middle ; the second in the apical third of the cellule. Abdomen thickly covered with long rufo-fulvous hairs ; the petiole black ; the sides with large deep punctures ; the keel blunt ; the second segment indistinctly punctured ; the last two segments black ; the apices of the others fringed with long orange hairs : the last segment is more strongly punctured ; and is stoutly keeled at the sides.

MUTILLA DILECTA, sp. nov.

Nigra, thorace rufo, mesopleuris nigris; alis fusco-hyalinis. ♂. Long. 8 mm.

Hab. Barrackpore (*Rothney*).

Head black, coarsely punctured, covered densely with white pubescence and more sparsely with long fuscous hairs; a furrow leads down to the antennæ ; head rounded and narrowed behind the eyes, the vertex less than half the length of the eyes ; an indistinct furrow at the sides of the ocelli; mandibles entirely black; palpi pale. Antennæ stout, the flagellum covered with pale down ; the scape sparsely haired. Prothorax large, red, except a somewhat triangular black mark on the lower part of the propleuræ : in front it is transverse ; the sides above straight, very slightly widened and straight towards the tegulæ. Mesonotum coarsely punctured, as also the scutellum ; the median segment strongly reticulated ; the reticulations large and all well defined, the central reticulation at the base being the largest, with the sides straight, and the apex triangular. The apex of the median segment is obliquely truncated. The propleuræ are coarsely longitudinally striolated ; the mesopleuræ somewhat strongly punctured except at the base and the apex, the latter being excavated, smooth, shining, and impunctate; the metapleuræ coarsely reticulated. The basal abdominal segment is a little longer than broad, strongly punctured, bearing long pale fuscous hairs, and, at the apex, pale golden hairs; the second segment is not quite so strongly punctured as the apex of the first; the other segments are less strongly, but more closely, punctured, than the second; the apical is more strongly punctured than the penultimate. The basal ventral segment is keeled down the middle. Legs entirely black, and covered with white hairs; the calcaria pale. Tegulæ densely covered with long white hairs. The basal abscissa of the radius straight, oblique; the apical roundly curved; the first recurrent nervure received in the middle; the second in the apical fourth of the cellule. The nervures are dark fuscous.

MUTILLA DISCRETA, *sp. nov.*

Nigra, longe argenteo-pilosa; alis fusco-violaceis. ♂. Long. fere 9 mm.

Hab. Barrackpore (*Rothney*).

Head densely covered with silvery pubescence; that between the ocelli and the antennæ completely hiding the texture; that on the vertex intermixed with long grey hair; vertex behind the eyes rounded and narrowed a little more than the length of the eyes. Antennal tubercles acute, piceous; the middle of the mandibles piceous. Antennæ black, the flagellum with a short pile; the scape with long hairs, and apparently more deeply excavated beneath than usual. Sides of the pronotum rounded, closely punctured and covered with long pale hairs. Mesonotum bearing large, round, deep punctures, the scutellum also with large deep punctures; at its base is a wide deep distinct furrow, behind which is a longer narrower one. The median segment has a gradually rounded slope, and is strongly reticulated.

Pleuræ strongly punctured; the apex of the propleuræ strongly crenulated; the base and apex of the mesopleuræ excavated, shining, impunctate, except on the lower part; the propleuræ reticulated, except at the base. The base of the abdomen excavated, projecting at the sides; the other segments punctured, but the punctation becoming weaker towards the apex; all the segments fringed with long white hairs. The keel on the basal ventral segment stout, a little curved, and a little projecting at the apex; the second segment has the punctures large and deep: the others have the base impunctate; the apex closely punctured, and with the oblique lateral furrows distinct; the apical half of the hypopygium roundly depressed. Legs densely covered with white hairs. The first transverse cubital nervure curved and bent at the lower third; the second sharply elbowed a little above the middle: both the recurrent nervures are received shortly beyond the middle of the cellule.

MUTILLA RUFODORSATA, *sp. nov.*

Nigra, dense argenteo-hirta; mesonoto rufo; abdomine nigro-cærulco; alis fusco-violaceis. ♂. Long. 13 mm.

Hab. Agra (*Rothney*).

Head narrower than the thorax, rugosely punctured, densely covered with silvery hairs all over; the mandibles before the tips piceous. Antennæ entirely black; the basal two joints covered thickly with silvery hairs; the flagellum with an indistinct down. Thorax densely covered with silvery hairs all over, black; the mesonotum and basal half of scutellum rufous. Pronotum and mesonotum coarsely punctured, almost reticulated; the scutellum very coarsely irregularly reticulated; the apex of the median segment has a sharp oblique slope; coarsely reticulated; the base thickly covered with silvery pubescence. The pleuræ coarsely punctured, except the base

E

and apex of the mesopleuræ; the metapleuræ coarsely reticulated. The legs are thickly covered with white hair. The basal two segments of the abdomen above are somewhat strongly punctured; the punctations becoming weaker towards their apices; all the segments at their apices are fringed with silvery hairs. The apical dorsal segment terminates in the middle in a triangular depression, with raised stout lateral keels, and with a central keel not half the width of the lateral ones. The second ventral segment is coarsely punctured: the sides at the base depressed, and with an indistinct keel between: the keel on the basal segment broad; the basal part the longest. Tegulæ large, rather strongly punctured. The wings are strongly fusco-violaceous, more lightly coloured at the base; the basal abscissa of the radius sharply oblique, the apical curved: the first transverse cubital nervure oblique, the second almost straight; the third sharply angled in the middle; the first recurrent nervure is received shortly before, the second at a slightly greater distance beyond the middle of the cellule.

MUTILLA PROVIDA, sp. nov.

Nigra, pro- meso-thorace mesonotoque rufis; alis fere fumatis. ♂. Long. fere 7 mm.

Hab. Bombay Presidency (*Wroughton*).

Comes very near to *M. dilecta*: but is easily known from it by the black scutellum and metathorax.

Head as wide as the mesothorax: entirely black, except the antennal tubercles, which are rufous: densely covered all over with long soft white hairs: rounded behind, where it is as long as the eyes: the palpi fuscous; the mandibles before the teeth rufous. Antennæ black; the scape with some white hairs: the flagellum with a fuscous down. Pro- and mesothorax coarsely punctured, covered with long white hairs: ferruginous, except the lower part of the propleuræ and the prosternum. Scutellum black; covered with long white hairs; the base and centre of the median segments with large, the sides with smaller reticulations, and having a gradually rounded slope. The propleuræ are coarsely punctured, almost reticulated; the mesopleuræ coarsely punctured and densely covered with long white hair; the upper part of the metapleuræ reticulated, the lower smooth. Legs black, densely covered with long white soft hairs; the calcaria pale. The alar nervures testaceous, slightly darker at the base: radial cellule wide; the basal abscissa of the radius oblique, straight, shorter than the apical which is curved, almost angled in the middle; the first transverse cubital nervure is oblique, slightly curved; the second is curved and bullated beneath hardly oblique; the third is sharply angled in the middle; the first recurrent nervure is received in the middle, the second in the apical third of the wing. Abdomen at the base rather strongly punctured, towards the apex, the punctation becomes weaker; covered, especially at the apices of the segments, with long white hairs. The basal abdominal segment coarsely punctured; the central keel, moderately strongly developed and hardly raised at the apex; the second segment strongly, the others much more weakly punctured: their apices fringed with long white hairs; the apical segment entire, not depressed, punctured throughout.

C. Species described from females only.

MUTILLA LUXURIOSA, *sp. nov.*

Nigra, thorace supra obscure ferrugineo; abdomine albo-sexmaculato. ♀. Long. 7—8 mm.

Hab. Ceylon (*Yerbury*).

Head not broader than the thorax; black; the mandibles in the middle, clypeus and the antennal tubercles rufous; coarsely punctured, covered with longish and white and fuscous hairs; behind the eyes less than their length; narrowed and rounded. Antennal scape covered with long pale hairs, not grooved; the flagellum obscure fuscous beneath; covered with an indistinct microscopic down; the third joint nearly twice the length of the fourth. Mandibles grooved on the outer side. Thorax a very little dilated gradually towards the apex; the pronotum coarsely shagreened; mesonotum coarsely rugosely punctured; median segment at the apex with an abruptly oblique slope; the propleuræ obscurely, the metapleuræ coarsely rugosely reticulated, its outer edge spinose; the mesopleuræ shining, a little excavated, smooth, shining, and glabrous. Legs thickly haired: the spines thick, pale and black on the tibiæ, rufous on the tarsi. Basal abdominal segment gradually dilated towards the apex, not distinctly separated from the second; on the latter are two oval whitish fulvous marks near the base; the third and fourth have two marks of the same colour; those on the latter the smaller; the apical is fringed laterally with pale long hairs, and is closely aciculated or shagreened. The basal ventral segment is obscure rufous; the keel in the middle has a longer and a shorter blunt tooth; the second segment has widely scattered punctures; the others are finely transversely striated at the base; the apex with scattered punctures and covered with long pale fulvous hairs.

Resembles closely *M. aulica* Sm., in coloration, but wants the large spot of silvery pubescence on the vertex, and otherwise is easily known from *M. aulica* by having the pronotum at the base transverse, with the sides acute; while *M. aulica* has the sides broadly narrowed and the base not transverse. In one of my examples of *M. aulica* there are only four white spots on the abdomen.

Mutilla remota, sp. nov.

Long. 15 mm. ♀.

Hab. Trincomali (*Yerbury*).

Comes very near to *M. egregia* Sauss., in the coloration of the abdomen and in the sides of the thorax having a stout spine; but differs in having the head and thorax entirely black, not red.

Head narrower than the thorax, deep black, coarsely rugosely punctured, thickly covered with long black, intermixed on the front with shorter golden hairs; the orbits on the outer side narrowly rufous. Antennal scape rufous, covered with long golden hairs. Thorax at the base narrower than the head, becoming gradually wider to the spines, then becoming rather abruptly narrower to the apex; coarsely rugosely punctured, sparsely covered with long black hairs; the apex of the median segment with an abrupt oblique slope; coarsely rugosely punctured; the hairs long; on the upper part black, on the lower golden. The pro- and meso-pleuræ rugosely punctured; the latter projecting in the middle and ending at the top in a stout rufous spine; the space beyond the spine a little hollowed and finely transversely striated; the metapleuræ entirely rugosely punctured. The legs black; the tibiæ rufous; the femora covered with long black hairs; the tibiæ and tarsi more thickly with long golden hairs. The petiole narrow at the base, gradually dilated to the apex; thickly covered with long golden fulvous hairs, broadly at the apex, narrowly at the base; the intermediate space covered with long black hairs. Second segment coarsely punctured; covered with short black hairs; the apex with a belt of golden pubescence; the fourth segment covered with golden pubescence; the other segments covered with long black hairs. The basal two ventral segments covered with fuscous; the third and fourth covered with golden pubescence.

MUTILLA MANDERSI, sp. nov.

Nigra, thorace rufo; abdomine fulvo-sexmaculato; basi femorum late rufo. ♀. Long. 17 mm.

Hab. Shan States (*Manders*).

Comes very close to *M. funaria*, but differs from it in its longer and, as compared with the head, somewhat narrower thorax, which is further entirely rufous, besides being less strongly punctured; and by the legs being broadly rufous at the base. It comes near also to *M. sexmaculata*, but that has the thorax more thickened towards the apex, with its sides entirely black, while the marks on the abdomen are much more elongated on the second segment.

Head as wide as the thorax, black, coarsely rugosely punctured, covered with fuscous and golden hairs: behind the eyes it is rounded, and is there somewhat shorter than the length of the eyes: the mandibles are rufous in the middle. Scape of the antennæ covered sparsely with long silvery hairs; the flagellum with an obscure down; the third joint twice the length of the fourth. Palpi black. Thorax more than twice the length of the head; the sides almost straight: hardly dilated towards the apex: above closely and strongly punctured; covered with longish fuscous hairs: the apex with a semi-abrupt slope, rounded at the top; the pleuræ smooth, shining, beneath covered with a silvery down; the base and apex obscurely punctured. Abdomen with the basal three segments as long as the head and thorax united: the top covered thickly with black: the sides and ventral surface more sparsely with longer silvery hairs; velvety: the basal segment gradually dilated towards the apex: the two marks on the second segment are oval, large: on the third they are more than twice broader than long; on the fourth they are not much longer than broad; there are none on the fifth; the pygidium is densely covered at the sides with long fulvo-aureous hairs. The basal ventral segment is coarsely punctured, and has in the middle a projection, which rises a little towards the apex, which is a very little curved; the second segment has distinctly separated punctures; the others are finely and closely punctured on the apical half and thickly clothed with long pale fulvous hairs. The legs are moderately pilose; the tibial spines fulvous or pale; the fore femora are rufous at the sides and beneath in the middle; the two hinder pairs rufous, except at the base and apex, where they are black.

In size and form it comes near to *M. sex-maculata*, but may be known from it by the thorax being entirely red; and by the third and fourth abdominal segments having interrupted white bands instead of spots; the marks on the second segment, too, being oval and not elongate.

MUTILLA VALIDA, *sp. nov.*

Nigra. thorace supra rufo, abdomine albo-bimaculato, basi longe fulvo-hirto. ♀. Long. 8 mm.

Hab. Barrackpore (*Rothney*).

Head slightly wider than the thorax; thickly covered with long, the sides more thickly with silvery pubescence; behind the eyes it is developed twice the length of the eyes. Scape covered with long white hairs; the flagellum with fuscous down. Thorax about one half longer than the head, its sides straight; the pronotum and the apical three-fourths of the median segment black; the rest reddish; strongly punctured, almost reticulated; the apex with an oblique slope; slightly hollowed, smooth; the pro- and meta-pleuræ coarsely reticulated, the mesopleuræ impunctate, smooth; the lower part thickly covered with white hairs. The basal segment of the abdomen smooth, obscurely shagreened, the apex with a broad, thick band of rich fulvous hair; the second segment with two oval marks of pale fulvous hairs; the apex with a broad fulvous belt of thick hairs; the third covered with thick hair, laterally pale, in the middle rufo-fulvous; the other segments fringed with pale fulvous hair. Ventral segments sparsely covered with pale hairs: the second strongly punctured.

Mutilla humilis, sp. nov.

Nigra; capite et thorace rufis; abdomine argenteo 4-maculato; pedibus anticis obscure rufis. ♀. Long. 4 mm.

Hab. Barrackpore (*Rothney*).

Head wider than the thorax, rufous: the orbits broadly black; the vertex and front obscurely longitudinally striolated; the oral region and the palpi testaceous; the mandibles piceous before the apex. Thorax above ferruginous, the sides all round the top, bordered with black and irregularly longitudinally striolated; the edges irregular, rough; with a few teeth which are more numerous on the sides of the median segment, which has a gradually rounded slope, and has at the top a large tooth. Pleuræ shining, impunctate, glabrous. Legs black; the anterior tibiæ and tarsi obscure rufo-testaceous. Abdomen black, sparsely covered with long black hairs; on basal segment is a square of silvery pubescence: on the apex of the second segment are three oval silvery marks; the other segments are marked with silvery pubescence in the middle.

Mutilla laeta, sp. nov.

Nigra, thorace supra rufo; abdomine argenteo 4-maculato. ♀. Long. 9 mm.

Hab. Barrackpore (*Rothney*).

Head hardly wider than the thorax, very coarsely rugosely punctured, covered with long glistening hair; a rufous mark in the centre of the vertex. Scape of the antennæ covered with long glistening hairs; the flagellum with a distinct down; the thorax is not quite twice the length of the head; the sides looked at from above are straight; but the mesopleuræ are slightly excavated; above coarsely uniformly punctured; the propleuræ entirely black, coarsely punctured; the mesopleuræ smooth shining, separated from the propleuræ by a keel; black, rufous above, the lower part densely covered with silvery pubescence; the metapleuræ coarsely reticulated, rufous above, black below. The median segment has at the apex a sharp oblique slope. Abdomen black, covered with black hairs, two elongated oval marks of silvery pubescence on the second segment near the middle; the third and fourth segments with silvery pubescence at the sides above: the other segments fringed with silvery pile. The ventral segments covered with silvery hairs; the second segment strongly punctured, the punctures distinctly separated. Legs entirely black, densely covered with stout hairs; the tibial spines stout, longish.

MUTILLA PUERILIS, *sp. nov.*

Ferruginea, sparse longe albo-hirta; abdomine nigro. ♀. Long. fere 8 mm.

Hab. Ceylon (Yerbury).

Antennæ ferruginous, the scape shining, sparsely covered with longish white hairs; the flagellum closely covered with short white down. Head shining, sparsely covered with longish white hairs; the tips of the mandibles and an elongated mark extending from the antennal tubercles to the eyes, becoming narrowed as it approaches them, black. Thorax above coarsely rugose, becoming somewhat reticulated towards the median segment; the mesothorax hardly narrowed towards the middle. The propleuræ a little shagreened; the base of the mesopleuræ a little aciculated and hollowed; the rest very shining and impunctate; the apical part of the metapleuræ with distinctly separated punctures. The median segment is rounded at the top; the apex semi-oblique; abdomen shining, closely and minutely punctured; sparsely covered with long pale hairs; the base of the first to third segments in the middle bearing long pale golden hairs; the hypopygium covered with long pale golden hairs and closely punctured; the ventral segments black, sparsely covered with long white hair; the second segment bearing large, widely separated punctures; the other segments more closely and finely punctured at the apex. Legs covered with long stiff whitish hairs; the anterior tibiæ and femora slightly infuscated.

MUTILLA ARIEL, *sp. nov.*

Ferruginea; abdomine cæruleo, argenteo-maculato; antennis pedibusque nigris; basi antennarum ferrugineo. ♀. Long. 11 mm.

Antennæ black, sparsely pilose; the basal two joints and the greater part of the third rufous. Head ferruginous; above the antennæ blackish; coarsely punctured; sparsely covered with long fuscous hairs. Thorax ferruginous, not half the length of the abdomen, the sides rounded at base and apex; the latter with an oblique slope, broadly rounded at the top; above coarsely punctured, sparsely covered with long blackish hairs, which become silvery white on the median segment, there is a large pale golden spot on the base of the mesonotum; the sides are slightly and gradually narrowed from the base of the mesothorax to the apex. The mesopleuræ a little hollowed in the centre, infuscated; the lower part densely covered with pale golden hair; the pro- and meta-pleuræ coarsely punctured. The abdomen is metallic blue, shining, sparsely covered with long black hairs; there is a spot of silvery pubescence on either side of the basal segment, an oblong or oval one in the middle of the second; the second segment at the apex has a broad band of silvery pubescence, dilated, broadly and roundly in the middle; the other segments above in the middle bear silvery spots; the ventral segments are somewhat thickly covered with long silvery pubescence. The legs are covered with longish stiff silvery hairs; the tibial spines stout: the calcaria pale testaceous.

Comes nearest to *M. regia* Sm., of which it may be a variety, but is larger judging by the examples at my disposal: the latter has the flagellum rufous, not black; the legs for the greater part rufous, and the thorax more dilated at the base. The head and thorax want the metallic brassy tint of *M. metallica* and *M. pulchriventris*.

MUTILLA DIVES, *sp. nov.*

Nigra: thorace supra ferrugineo; abdomine argenteomaculato. ♀. Long. 8 mm.

Hab. Barrackpore (*Rothney*).

Antennæ entirely black, stout, as long as the head and thorax united. Head a very little narrower than the thorax, coarsely punctured, sparsely covered with fuscous hairs, thorax coarsely rugosely punctured; the base rounded, the sides not contracted as seen from above; the mesopleuræ excavated, shining, impunctate; densely covered with long silvery hairs. The median segment sharply oblique; the metapleuræ coarsely rugose; the propleuræ with shallow punctures. Abdomen a little longer than the head and thorax united, deep black, velvety, the base of the second segment with silvery pubescence in the middle; a somewhat roundish spot above in the middle, and a slightly smaller one at the apex; the third segment is covered entirely with silvery pubescence; and the apical segment is fringed with long silvery hairs. Legs entirely black, densely covered with long white hairs.

Is not unlike *M. taprobanæ* but is longer, and has the pleuræ entirely black.

Mutilla peregrina, sp. nov.

Long. 7 mm.

Hab. Barrackpore (*Rothney*).

A smaller and more slender species than *M. discreta*, from which it may be known by the mesonotum being more coarsely punctured and having besides two stout longitudinal furrows.

Head distinctly narrower than the thorax; the part behind the eyes, almost less than their length; strongly punctured; thickly covered with long white hairs; the mandibles ferruginous before the teeth; the palpi dark fuscous. Antennæ stout; the scape grooved beneath; sparsely covered with long pale hair; the flagellum thickly covered with fuscous down. Pronotum strongly and closely punctured: the mesonotum strongly punctured; and with the punctures more widely separated; the longitudinal furrows wide and continuous; the scutellum less strongly punctured than the mesonotum; the median segment with a gradually rounded slope; strongly reticulated. Pleuræ coarsely punctured; the base of the metapleuræ impunctate, the apex reticulated. The basal abdominal segment very coarsely punctured; the second less strongly, the other segments almost impunctate, and rather densely covered with long white hairs. The basal ventral segment coarsely punctured, without a keel; the second rather strongly punctured; the others finely and closely punctured at the apex; the apical segment closely punctured, without any depression.

Mutilla Cotesii, sp. nov.

Nigra; thorace capiteque argenteo-maculatis; pedibus nigris, basi rufis. ♀. Long. 8 mm.

Hab. Barrackpore (*Rothney*).

Head large, wider than the thorax, but not much; black; a large somewhat roundish mark of silvery pubescence on the centre of the vertex, the edge of the occiput rufous; the vertex strongly longitudinally striolated; the front strongly striolated; the antennal tubercles shining, rufous; the middle of the mandibles broadly rufous. Antennæ covered with white pubescence; the flagellum for the greater part rufous beneath. Thorax not one-half longer than the head; very slightly widened towards the apex; the edges at the top irregular; those of the median segment with four large pale rufous teeth, the apex of the median segment oblique, but not sharply The pleuræ shining, impunctate, black, except the apices of the metapleuræ which are rufous; the mesopleuræ are hollowed at base and apex. Legs black, covered with white hair; the tibial spines long, pale rufous. The coxæ, trochanters, and base of femora (the hinder broadly) rufous. Abdomen not much longer than the head and thorax united; the basal segment at the base obliquely truncated, and at the apex distinctly separated from the second; and with a square mark of silvery pubescence in the centre at the apex; the second segment has at the apex three somewhat oval marks of silvery pubescence; the fourth and fifth segments have silvery pubescence at the apex. The keel on the basal ventral segment ends in the centre in two teeth, the basal being twice the length of the apical; the second segment bears large, round deep punctures; the centre is a little raised, and the raised part ends before the apex in a blunt raised, somewhat triangular tooth. The other segments are more closely punctured, except at the base; and all are covered with long pale soft hairs.

MUTILLA ROTHNEYI, *sp. nov.* (Pl. 4, f. 14).

Capite thoraceque supra ferrugineis, abdomine nigro, argenteo 5-maculato. ♀. Long. 8 mm.

Hab. Barrackpore (*Rothney*).

In coloration and form of the head and thorax like

M. œdipus, but the latter is easily known from it by the thorax being contracted in the middle.

Head large, a little broader than the thorax: coarsely longitudinally striolated; the striæ running into reticulations towards the antennæ; black; the front and vertex broadly ferruginous; the ferruginous colour extending to a little below the bottom of the eyes: the antennal tubercles and a stripe on the mandibles ferruginous. The scape of the antennæ covered with long silvery hairs, the flagellum sparsely with a pale down. Thorax short, not much longer than the head; coarsely rugosely punctured, sparsely covered with black hairs; the apex of the median segment oblique: the pleuræ excavated, shining, impunctate; the base and apex a little pilose: at the top in the middle the median segment ends in a sharp spine. Abdomen a little longer than the head and thorax united; the base obliquely truncated, with a narrow margin at its apex, and with a spot of pale golden pubescence in its centre above; at the apex of the second segment there is a central and a lateral somewhat larger oval mark of pale golden pubescence; the other segments have a somewhat squarish mark in the middle of the same colour. The ventral segments punctured, the basal segment much more strongly than the others; they are fringed with fulvous hairs. Legs entirely black, bearing white hair.

The present species may, of course, be the ♀ of my *M. œdipus*, but this is a point which can only be decided by direct observation. The head in *M. Rothneyi* is wider compared to the prothorax; the mesothorax is stoutly spined; the front and vertex strongly longitudinally striated all over, while in *M. œdipus* it is only punctured; the head behind the eyes in *M. œdipus* is much more thickly covered with white hairs. There is no appreciable difference in the form and coloration of the two; in *M. Rothneyi* the sides of the median segment are stoutly spined all over; in *M. œdipus* there are only five large, stout, widely separated spines, these becoming larger from the base to the apex; in *M. Rothneyi* the second basal abdominal segment is stoutly produced in the middle towards the apex, which is not the case in *M. œdipus*.

FOSSORES.

OXYBELUS CEYLONICUS, *sp. nov.*

Long. 7 mm.

Hab. Trincomali, Ceylon (*Yerbury*).

Approaches nearest to *O. squamosus* Sm., with which it agrees best in the form of the squama; but *O. squamosus* may be known from it by the hinder tibiæ and tarsi being yellow, while here they are black.

Head closely punctured, covered, with short white pubescence, especially above, where it assumes a fulvous hue. The scape of the antennæ black above, yellowish beneath; the flagellum fulvous; its base yellowish, blackish above. The clypeus projects and is thickly covered with longish silvery hairs; the mandibles yellowish at the base, black at the apex, and piceous between. Thorax black; a line on the pronotum behind; the tegulæ, tubercles, sides of the scutellum, and the squamæ on the post-scutellum yellow. Mesonotum closely punctured, the punctures more widely separated towards its apex; the suture at the base of the scutellum crenulated; the scutellar punctures large, widely separated, more numerous at the apex, where there is, in the middle, a stout projection. Post scutellar squamæ curved on the outer side, ending in a curved triangular tooth; the squama large, curved at the base, where there is a stout longitudinal keel; the rest of it with stout striations all clearly separated; the apex roundly incised; the ends rounded; the segment at the side of the squama is smooth except for a few striations, and is for the most part pale brownish; its outer side aciculated and with a few stout striations: the apex has in the middle two stout keels, which form a large triangular space rough in the centre and depressed at the apex: the keels prolonged as one to the apex: the sides obliquely aciculated. The mesopleuræ punctured, the metapleuræ obliquely striated: the striations widely separated. Legs: the coxæ and the base of the fore femora black; the fore femora yellowish; the four posterior ferruginous, the fore tibiæ yellowish; the middle ferruginous; the hinder blackish; the tarsi blackish; the anterior testaceous, yellowish at the base: the hind tibiæ strongly spined, the spines white, the spurs of a more testaceous hue; the apex of the middle femora and the four hinder tibiæ at the base, yellow. Wings clear hyaline, the nervures dark fuscous. Abdomen black, strongly punctured, the sides with a broad yellow line on the four basal segments: the pygidial area thickly covered with longish fulvous hairs.

Astata tarda, sp. nov.

Nigra; abdomine rufo late balteato; alis hyalinis, fere fumatis; stigmate testaceo. ♂. Long. 10 mm.

Hab. Ceylon (*Yerbury*).

Resembles *A. agilis* Sm., but is larger, and has not the wings distinctly smoky at the apex; the radial cellule at the top much longer than the stigma, while in *A. agilis* it is only about its length: the third cubital cellule much longer compared to the second: and the median segments with no distinct longitudinal keels. Head closely punctured: the sides and clypeus thickly covered with long white hairs; a furrow leads down to the antennæ, the space between the latter smooth and shining, as is also the space in front of the ocelli. Scape of the antennæ covered with long white hairs.

Thorax covered with long white hairs ; the mesonotum closely punctured ; the scutellum at the base smooth, impunctate ; post-scutellum rugose ; the median segment reticulated ; its top on the oblique apex with a deep oval, impunctate depression ; a wide deep oblique depression on the mesopleura. The basal segment of the abdomen black ; its apex, the second segment and the base of the third, ferruginous ; the base and apex thickly covered with long white hair ; the basal ventral segment ferruginous at the sides.

PISON (PARAPISON) ROTHNEYI, *sp. nov.* (Pl. 4, f. 15).

Nigrum, argenteo-pilosum ; mandibulis geniculis, tibiisque anticis rufis ; alis hyalinis. ♀. Long. 6—7 mm.

Hab. Barrackpore (*Rothney*).

Black, shining, covered with silvery pubescence. Head finely punctured, covered with short fuscous pubescence, except at the incision of the eyes ; the cheeks thickly and the clypeus more sparsely covered with silvery pubescence ; an indistinct shallow longitudinal furrow below the ocelli ; the mandibles and palpi rufous. Antennæ black ; the apical three joints rufous beneath ; covered with sparse fuscous pubescence. Thorax shining, impunctate ; sparsely covered with white pubescence, especially at the sides and base ; the median segment with a broad longitudinal furrow extending from the base to the middle of the segment. Pleuræ sparsely covered with short white hairs ; the apex of the propleuræ brownish and surrounded by a fringe of silvery pubescence ; in the centre of the mesopleura is a deep short depression, a little longer than wide ; the sternum deeply and widely excavated. Legs black, thickly covered with short silvery pubescence ; the apical third of the fore femora, the front tibiæ entirely, the hinder side of the middle, the basal three-fourths of the hinder, the hinder knees and the calcaria, rufo-

F

testaceous. Wings hyaline, the costa and nervures black, the latter paler towards the apex of the wing; the second cubital cellule is much narrowed towards the top, the space there bounded by the transverse cubital nervures being less than that bounded by the first recurrent and the first transverse cubital nervure; the second recurrent nervure is almost interstitial nervure. Abdomen shining, impunctate; sparsely covered towards the apex with a silvery pile; the apical segments at the apices testaceous.

Parapison was erected by Smith (*Trans. Ent. Soc.*, *1869, p. 298*) for those species, otherwise agreeing with *Pison*, which have only two transverse cubital nervures. Kohl (Die Gattungen und Arten der Larriden, *Verh. z.-b. Ges. Wien, xxxiv.*) regards it as only a section of *Pison*.

Pison striolatum, sp. nov.

Nigrum; facie argenteo-pilosa; alis hyalinis. ♀. Long. 8 mm.

Hab. Mussouri (*Rothney*).

Has the typical neuration of *Pison* as figured by Kohl, (*Verh. z.-b. Ges. Wien., xxxiv., t. 8, f. 1,*) i.e., the recurrent nervures are both interstitial.

Head in front opaque, coarsely rugose, behind the ocelli the vertex more shining, and with the punctures more distinct and much more widely separated; below the centre of the eye incision thickly covered with silvery pubescence; the apex of the clypeus gradually brought to a sharp point. Thorax black; sparsely covered with fuscous, the median segment with longer white hairs; the sides with oblique, the centre with curved striæ; the base with a short straight keel at the base; the apex is broadly depressed. Pleuræ strongly punctured: a wide longitudinal furrow on the mesopleura; the metapleuræ smooth; covered sparsely with long hair. Tibiæ and tarsi thickly covered with white pubescence, which gives them a whitish appearance. The two recurrent nervures are completely interstitial; the pedicle of the second recurrent nervure is longer than the lower cellule. Abdomen smooth, shining, impunctate; the apex sparsely covered with white hairs.

CEMONUS.

Neither this genus (or subgenus according to some authors) nor its type *Pemphredon* has been recorded hitherto from the Oriental Region.

CEMONUS FUSCIPENNIS, *sp. nov.* (Pl. 4, f. 16).

Niger, nitidus, sparse albo-hirsutus; alis fumatis, basi fere hyalinis. ♀. Long. 8–9 mm.

Hab. Mussouri (*Rothney*).

Head very shining, the front closely, the vertex much more sparsely covered with shallow punctures; sparsely covered with long fuscous hairs; the cheeks and clypeus at the sides much more thickly covered with long white hairs; the clypeus with a few punctures in the middle; mandibles entirely black. Antennæ entirely black; almost bare. Thorax shining; the pronotum coarsely punctured; the mesonotum shining, smooth: in front with a few widely separated punctures: and an indistinct, shallow, longitudinal furrow; and there is a more distinct lateral one. The smooth area in the base of the median segment is stoutly crenulated; and there is a distinct longish longitudinal furrow in the centre: and the sides (but not at their extreme edges) have some shallow punctures. The propleuræ at the base are strongly aciculated; the centre smooth, almost impunctate, the mesopleuræ strongly punctured; the metapleuræ obliquely striolated, almost reticulated. Legs black; the femora and tibiæ sparsely covered with long white hair; the tibiæ also with a pale pubescence; the tarsi have the hairs thicker and shorter. Wings, nervures, and stigma black: the second recurrent nervure received shortly before the first transverse cubital. Petiole rugosely punctured; covered with long white hair; the rest of the abdomen very smooth and shining, almost glabrous, except at the apex, where there are a few pale hairs.

POMPILIDÆ.

CEROPALES ALBOVARIEGATA, sp. nov.

Lutea: capite thoraceque albo-maculatis; vertice, pronoto basique mesonoti, nigris; alis hyalinis. ♂. Long. 8 mm.

Hab. Trincomali, Ceylon (*Yerbury*).

Antennæ black; the basal two joints of the flagellum brownish beneath; the scape yellowish beneath; the flagellum thickly covered with short pubescence. Head shining, impunctate; the front, vertex and occiput, except at the sides, black: the rest white; the labrum brownish; the tips of the mandibles black; the antennal tubercles largely projecting above the antennæ, clear white; oval, deeply triangularly cleft down the middle: the anterior ocellus in a depression from which runs a furrow. Thorax smooth, shining, impunctate, glabrous; the pronotum black, lined with white at the apex; its sides at the base projecting beneath, clear white; the mesonotum black to the scutellum, in front of which it is white, it having also a white line at the tegulæ; there is a white mark under the tegulæ, a smaller one under the hind wings; a large mark at the base of the mesopleuræ on the lowerside, and a narrow line opposite it at the apex, a small triangular oblique mark and a curved one over the hind coxæ, clear white: legs fulvous: the anterior coxæ white beneath; the hind tarsi fuscous. Wings clear hyaline; the stigma and nervures black.

POMPILUS ICHNEUMONIFORMIS, *sp. nov.*

Fulvus; capite, thorace apiceque antennarum nigro-maculatis; alis fulvo-hyalinis, stigmate fusco. ♀. Long. 13—14 mm.

Hab. Mussouri (*Rothney*).

Head fulvous; a broad black band with straight sides, extending from the ocelli to near the eyes; the vertex sparsely covered with fuscous; the clypeus much more densely with longer fulvous, hairs; the apex of the clypeus transverse, smooth, and shining; the orbits have a yellower tint than the clypeus; mandibles yellow, black at the apex. Antennæ fulvous; the apical five joints black; thorax fulvous; the base of the pronotum, of the mesonotum, two broad black lines on it extending from the base to the scutellum, the sides at the post-scutellum, the base of the median segment, the pleural sutures and the sternum, black. Legs entirely fulvous; the coxæ with a more yellowish hue. Wings fulvous, the apex with a smoky violaceous hue: the second cubital cellule at the top is a little longer, at bottom shorter than the third; the first recurrent nervure is received shortly before, the second in, the basal third of the cellule. Abdomen fulvous; the extreme base black; where there is a deep triangular depression.

DIPLOPTERA.

RHYNCHIUM BASIMACULA, *sp. nov.* (Pl. 4, f. 17.)

Nigrum; facie abdomineque flavo-lineatis; alis violaceis. Long. 16 mm. ♀.

Hab. Barrackpore (*Rothney*).

Black; the apex of the first and of the second abdominal segment much more narrowly, and a short line behind the eyes, white; the wings violaceous. Front and vertex with large, clearly separated punctures: the clypeus with the punctures obscure; its apex depressed; slightly curved: the base of the clypeus, and the inner orbits to the top of the incision, white; the mandibles deeply grooved. Antennæ entirely black. Thorax opaque; black: the edge of the pronotum white behind; the pro- and meso-notum strongly, but not very deeply punctured. Scutellum and post-scutellum more rugosely punctured than the mesonotum. Pro- and meso-notum more strongly punctured than the metapleuræ, of which the punctures are more widely separated and not so deep. Legs black, covered with a white down. Abdomen shining, impunctate; the apex of the basal segment with a broad white band of equal width; the second with a much narrower band, dilated slightly at the sides. Wings violaceous: the second cubital cellule at the top is as long as the space bounded by the second recurrent and the second transverse cubital nervures.

The ♂ is very similar to the ♀; the clypeus is entirely pale yellow, with a mushroom-like black mark at the apex; the line on the orbits is broader; the scape is for the greater part yellow, beneath.

This species was recognised by the late Mr. F. Smith as an undescribed species; but it has not, so far as I know, been described.

EUMENES BUDDHA, sp. nov.

Nigra: clypeo, lineis pronoti, tibiisque anterioribus flavis; alis fusco-violaceis. ♀. Long. 15 mm.

Hab. Barrackpore (*Rothney*).

Head black; the clypeus, and a narrow line dilated at the top reaching from it to near the ocelli, lemon yellow; the clypeus triangularly cleft in the middle at the apex forming two straight teeth; the part behind these being oblique; the teeth and apex of the mandibles rufous; front and vertex strongly punctured, sparsely covered with white down; there is a short yellow line behind, and close to, the eyes above the middle. Antennæ black, except the terminal hook and the two joints in front of it which are fulvous; the hook sharply curved. Thorax coarsely punctured all over, sparsely covered with white pubescence; a very narrow line on the side of the propleuræ joined to a larger one on the mesopleura, a line on the apex of the prothorax, one in the centre of the post-scutellum and a line on the apex of the first and second abdominal segments, lemon-yellow. Propleuræ shining, impunctate, deeply obliquely excavated; there is a short longitudinal furrow in the centre of the mesopleura; the oblique space below the hind wing smooth shining, impunctate, except the lower half at the base, which is strongly punctured; the upper edge at the apex crenulated. The median segments at the apex roundly depressed. Legs black, covered with a white down; the hair on the under side of the front tarsi is fulvous.

EUMENES ADVENA, *sp. nov.*

Nigra; tegulis pedibusque rufis; alis fusco-violaceis. Long. 15 mm. ♀.

Hab. Barrackpore (*Rothney*).

Antennæ entirely black; the scape shining, the flagellum coarse, opaque. Punctures of the head, close, especially on the front where they run into reticulations; the clypeus very thickly, the rest of the head more sparsely covered with silvery hairs; the front somewhat triangularly produced between the antennæ. Mandibles with two deep, wide longitudinal grooves on the apical half, the grooves being piceous towards the apex. Thorax coarsely punctured, more sparsely in the middle of the mesonotum, and still more sparsely on the scutellum; the parapsidal grooves are deep, wide, and run from base to apex. The propleuræ are coarsely irregularly obliquely striolated, on the lower half the striations being stout: the rest of it, being irregularly punctured and on the lower part irregularly striolated; the mesopleuræ strongly punctured; the punctures widely separated; shortly beyond the centre is a wide deep depression bearing some stout irregular keels; the hinder part of the mesopleuræ strongly aciculated. Scutellum shining; the middle at the apex a little depressed; at the base is a wide depression with five stout keels; and at its apex is a deep wide depression with oblique median segment depressed in the middle; the sides of the depression oblique and meeting at the bottom, and irregularly, somewhat obliquely striolated, the bottom and sides with a sharp keel; the lower part outside this central depression strongly irregularly reticulated. Legs red, covered sparsely with white hairs; the greater part of the fore coxæ and the base of the middle pair above, black. Abdomen black; covered with white hairs; a long depression, keeled in the centre, down the middle at the base; the second segment punctured; the other segments also punctured, but with the punctures closer together; and covered with long white hair.

ANTHOPHILA.
ANDRENIDÆ.

As will be seen, the Andrenidæ are almost exclusively confined to the northern parts, and more particularly to the mountainous regions of India.

PROSOPIS.

Only one species of *Prosopis* has been recorded from the Indian Peninsula; but several are known from the Australian portion of the Malay Archipelago. The new species here described may be separated as follows:—

1 (2) Front strongly longitudinally striated (only the hinder tibiæ yellow at the base). *striatifrons*

2 (1) Front not longitudinally striated.
3 (4) Tibiæ not annulated with white. *leucotarsis*
4 (3) Tibiæ annulated with white at the base.
5 (6) The keel on the median segment rounded inwardly in the middle at the apex and not running into a central furrow. *strenua*
6 (5) The keel on the median segment not rounded inwardly in the middle at the apex.
7 (8) The middle of the median segment raised and separated from the sides, which are not striated, base of tibiæ black. *obsoleta*
8 (7) The middle of the median segment not raised and separated from the sides and uniformly rugose; base of tibiæ white. *bellicosa*

PROSOPIS STRIATIFRONS, *sp. nov.*

Nigra; orbitis oculorum infra lineaque pronoti flavis; alis hyalinis. ♀. Long. 7 mm.

Hab. Barrackpore (*Rothney*).

Comes near to *P. obsoleta;* but is larger and stouter; the front is stoutly longitudinally striated; the median segment longitudinally striated.

Head black; the inner orbits yellow, the yellow dilated towards the middle, gradually at the bottom, more sharply and obliquely at the top. The front and clypeus longitudinally striated; the front raised, its sides with a stout border; the vertex finely punctured, very sparsely and shortly haired. Antennæ stout, the flagellum obscure brownish beneath. Thorax closely punctured, the mesonotum rather strongly punctured; the scutellum with the punctures finer and more widely separated. Centre of median segment at the base irregularly and somewhat strongly reticulated; this reticulated part being surrounded by a distinct border; the rest of the segment coarsely aciculated; the apex with a furrow down the centre. Mesopleurae more strongly punctured than the mesonotum: the oblique furrow obscurely crenulated. In the front of the pronotum is an interrupted yellow line; the tegulae are yellow in front, fuscous behind, and below them is a yellow mark. Wings hyaline; the costa and stigma black; the nervures more fuscous in tint: the first recurrent nervure is received shortly in front of the transverse, almost interstitial; the second interstitial. Legs covered with white hairs; the fore tibiae broadly obscure rufous (perhaps discoloured) in front; the basal third is lined with yellow behind; the base of the hind tibiae clear yellow. Abdomen shining, glabrous, impunctate.

PROSOPIS LEUCOTARSIS, *sp. nov.*

Nigra: clypeo, tegulis tarsisque flavis; alis hyalinis, nervis stigmateque pallidis. ♀. Long. 5 mm.

Hab. Ceylon (*Rothney*).

Head nearly as wide as the thorax: shining, impunctate; the front and vertex sparsely covered with long pale hairs; the clypeus and labrum yellowish-white, immaculate. Antennae black, sparsely microscopically pilose; the scape sparsely covered with fuscous hair. Thorax black, shining, bearing longish white hairs; the base of the median segment somewhat flat, aciculate; the apex with an oblique slope; covered with long pale hairs. Pleurae shining, impunctate, thickly covered with longish pale hairs, and having a bronzy tinge. Legs black, with a greenish tinge, sparsely haired; the tarsi rufo-testaceous, thickly covered with white hairs. The second recurrent nervure is straight, oblique; the second is, at the top, curved towards the first; the cellule at the top being there not much more than half the length it is at the bottom; the first recurrent nervure is almost interstitial; the second recurrent nervure is received very shortly in front of the second transverse cubital.

PROSOPIS STRENUA, *sp. nov.*

Long. 6 mm. ♀.

Hab. Barrackpore (*Rothney*).

Head as wide as the thorax, closely, but not very strongly, punctured; sparsely covered with pale hairs; a square mark on the clypeus, a line on the inner orbits extending from the base of the mandibles to near the level of the front ocellus, and becoming gradually wider in the middle, bright yellow; the labrum and tips of the mandibles piceous; the palpi testaceous. The front raised and stoutly keeled laterally over the antennæ. Antennæ covered with a fuscous pubescence; the base of the scape and the second and third joints, obscure rufous; the flagellum obscure testaceous beneath. Thorax black; a line on the base of the pronotum, narrowed in front, the tegulæ at the base and a mark in front of them, lemon-yellow. Mesonotum closely and rather strongly punctured; the scutellum with the punctures, if anything, larger, but more widely separated. Middle of the median segment raised, the raised parts forming a rugose triangle, bordered by a smooth space semicircular at the apex, which again is bordered by a distinct keel; its apex has a straight abrupt slope; the base in the middle excavated; the apex transversely rugose. Pleuræ rather strongly punctured; a wide furrow running down from the tegulæ to the sternum, which is strongly punctured and slightly hollowed in the middle. Legs black, slightly covered with white pile; the anterior greater part of the fore tibiæ, the base of the middle, the basal half of the hinder, and the greater part of the basal joint of the hinder tarsi, bright yellow; the front four tarsi testaceous, obscure yellow at the base, black at base behind. Wings hyaline; the nervures fuscous, the stigma and costa darker; the two recurrent nervures interstitial. Abdomen black, shining, the ventral segments obscurely punctured at the base; the apex bearing fuscous hairs.

PROSOPIS ABSOLUTA, *sp. nov.*

Long. 5 mm. ♀.

Hab. Barrackpore (*Rothney*).

Resembles *P. strenua:* but may be known by the base of the median segment wanting the smooth space and the curved keels.

Head scarcely so wide as the thorax, strongly punctured, almost rugose; thickly covered with short white hairs; a mark rounded at the base and broader than long at the apex of the clypeus; an elongate somewhat triangular spot, extending from the base of the clypeus to near the front ocellus, the apex of its dilated part being at the top of the clypeus, and the lines are united there by a yellow mark, which is truncated at the bottom, rounded at the top. Flagellum brownish beneath, covered with a microscopic down. Thorax black, strongly punctured; covered with short white pubescence; the median segment broadly and coarsely rugose in the middle, the rugose part triangular, and at the apex bearing stout transverse irregular keels; the sides and apex of the median segment alutaceous, covered with a pale microscopic down. Mesopleuræ rather strongly punctured, the punctures all distinctly separated; the metapleuræ coarsely alutaceous. Legs covered with white pubescence; the greater part of the fore tibiæ and the hinder four broadly white at the base; the front pair piceous behind; the calcaria white; the tips of the tarsi testaceous. Wings hyaline, the stigma and nervures dark fuscous; the first recurrent nervure received in front of the transverse cubital; the second interstitial. Abdomen entirely black: a spot of white pubescence on the side of the second segment.

PROSOPIS BELLICOSA, *sp. nov.*

Long. 6 mm.

Hab. Barrackpore (*Rothney*).

May be known from the other species by the central

part of the median segment being distinctly raised and separated from the lateral; these are smooth, shining, and impunctate; the surrounding keel is stout and piceous in colour.

Antennæ black; the flagellum brownish beneath, and bearing a slight white microscopic down. Head coarsely punctured; bearing short white microscopic pubescence; the inner orbits to near the lower ocellus, and dilated below the antennæ and narrower at the bottom than at the top, and a large mark on the clypeus, broad at the base and gradually narrowed to the top, yellow; the upper part of the mandibles yellow, the apical piceous. Thorax black; the pro- and meso-notum punctured; the median segment at the base with a large somewhat square coarsely rugose space, surrounded by a smooth impunctate area, bordered by a stout, semicircular piceous keel; the apex without a distinct furrow and thickly covered with white hairs. Pleuræ and sternum strongly punctured; a line on the pronotum at the apex, the tubercles and the base of the tegulæ yellow, the apex of the tegulæ, piceous. Legs black, covered with white pubescence; the anterior knees, tibiæ and tarsi, the middle tibiæ broadly at the base and slightly at the apex, the hinder tibiæ broadly at the apex and the metatarsi, yellow; the hinder femora incline to piceous beneath; and the yellow is suffused at the base or apex with brownish. Abdomen shining, the sides sparsely covered with white pubescence.

The size of the yellow mark on the clypeus varies.

HALICTUS.

a. Species with the abdomen more or less reddish.

HALICTUS WROUGHTONI, *sp. nov.*

Niger; longe dense pallide hirtus; abdominis basi late rufo; pedibus nigris, femoribus fere piceis, longe albo-pilosis; alis hyalinis, stigmate piceo, nervis pallidis. ♀. Long. 8mm.

Hab. Bombay Presidency (*Wroughton*).

Head black, very densely covered with grey pubescence, longer and slightly sparser in front; short and very dense behind, completely hiding the surface; the labrum fringed with long golden pubescence; the mandibles ferruginous at the apex, their base, on the outer side, covered with short, close, white pubescence. Antennæ black, slender, the scape sparsely covered with short white hairs. Thorax densely covered with soft white hairs; the mesonotum closely but not deeply punctured; the hairs on the postscutellum shorter and covering it entirely; the base of the median segment closely, irregularly striolated; the hairs on the pleuræ long and thick. Legs black; the femora dark piceous; covered with long white hairs, those on the hinder femora being especially long; on the hinder tibiæ and tarsi they have a fulvous tinge. The third cubital cellule is nearly twice the length of the second at top and bottom. Abdomen shining, impunctate; the basal segment and the basal half of the second red; the basal segment at the base covered with long white hairs; the others fringed with white pubescence; the ventral segments red; the apical two black; thickly covered with long pale hairs.

Halictus decorus, sp. nov.

Niger; abdominis basi late rufo; tarsis testaceis; alis hyalinis. ♀. Long. fere 5 mm.

Hab. Mussouri (*Rothney*).

Head with a slight greenish tinge; the clypeus and the lower part of the front thickly covered with white pubescence; the vertex with longer, sparser fuscous hairs; the mandibles ferruginous, the teeth blackish; their base sparsely covered with pale hairs. Antennæ black, brownish beneath from the fifth joint. Thorax black, almost impunctate; the pronotum thickly covered with white hairs; as is also the post-scutellum and the sides of the scutellum; the median segment has an elongated semicircular area, the base raised, rugosely punctured; the rest of it smooth, with a distinct central and a few incomplete longitudinal keels, there being also a transverse keel before the apex. Behind the narrow part of this area, at the sides of it, is a strongly irregularly obliquely keeled area; the apex of the segment semi-oblique, strongly aciculated. Meso- and meta-pleuræ thickly covered with white hairs; the latter obliquely striated. Legs black, with a piceous tinge, the femora darker; the femora sparsely, the tibiæ and tarsi thickly, covered with white hairs. Tegulæ testaceous, darker behind. Wings clear hyaline, the nervures and stigma blackish; the recurrent nervures received near the apical third of the cellules. Abdomen shining, smooth: the basal three segments red above and below; the apical black, thickly covered with longish white hair.

Comes near to *H. xanthognathus* Sm., but is much smaller and otherwise quite distinct.

HALICTUS DISSIMILANDUS, *sp. nov*.

Long. fere 6 mm.   ♀.

Hab. Mussouri (*Rothney*).

Is very like *H. decorus* in size and coloration, but may be at once known by the base of the median segment not having a depressed semicircular area clearly defined by a keel, it being instead strongly reticulated; the legs, too, are much lighter in tint.

Antennæ black, thick, sparsely covered with white microscopic hairs; the front and vertex more sparsely with longer hairs; the vertex obscurely punctured; mandibles rufous, the tips black. Pro- and meso-notum covered with fuscous hairs; the hairs on the scutellum longer and paler; the pro- and meso-notum coarsely punctured; the scutellum not quite so strongly punctured as the mesonotum; the median segment stoutly reticulated; its apex oblique shining, not reticulated. Propleuræ smooth, piceous; the base with some longitudinal keels; the apex fringed with woolly hairs; the mesopleuræ coarsely rugosely punctured; the metapleuræ with an oblique smooth space in the centre, the base and apex coarsely punctured. Legs obscure piceous; the femora darker; sparsely covered with white hairs. Wings hyaline; the nervures and stigma fuscous; the first recurrent nervure received quite close to the transverse cubital; the second in the apical third of the cellule. The basal three segments of the abdomen rufous, widely suffused with black in the middle; the apical segments sparsely covered with white hairs; the ventral segments coloured like the dorsal.

One of this species has projecting from the apex of the third dorsal segment the larvæ of one of the *Stylopidæ*, probably a *Halictophagus*.

HALICTUS INVIDUS, sp. nov.

Niger; thorace fortiter punctato; metanoto reticulato; abdominis medio ferrugineo; alis hyalinis, apice fere fumatis.
♀. Long. 7—8 mm.

Hab. Mussouri (*Rothney*).

Head rugosely punctured; below the antennæ thickly covered with white hairs; the front and vertex more sparsely covered with longer white hairs. Mandibles piceous in the middle; underneath with a few long golden hairs. Antennæ stout, the flagellum thickly covered with longish white hairs; the joints of the flagellum dilated broadly beneath; towards the apex bearing a white microscopic pile. Thorax coarsely rugosely punctured; the punctures on the apex of the mesonotum larger and more widely separated; the base of the median segment with stout longitudinal keels, irregular in the middle, forming almost reticulations; at the apex in the middle is an area broad and rounded at the base, becoming narrowed towards the apex; forming an almost pyriform space; there is an oblique, somewhat similar, area at its side at the top; the rest has four stout slightly oblique keels, running from the centre. Pro- and meso-pleuræ coarsely strongly punctured, running into strong reticulalations at the apex; the metapleuræ coarsely strongly punctured; the base coarsely obliquely striated; the punctures at the apex large, round. Legs black; the apices testaceous; the femora sparsely, the tibiæ, and especially the tarsi covered with silvery white pubescence. The first and second transverse cubital nervures are bullated at top and bottom; the first recurrent nervure is received quite close to the transverse cubital; the second shortly before the apical third of the cellule. Abdomen shining, the basal three segments punctured; the apical smooth and shining; the ventral segments shining, broadly ferruginous in the middle; the third ferruginous in the middle.

HALICTUS SERENUS, *sp. nov.*

Niger; abdominis basi late rufo: alis hyalinis, nervis testaceis. ♂. Long. 5 mm.

Hab. Mussouri (*Rothney*).

Head finely punctured, thickly covered with long white hair; the mandibles before the apex ferruginous; beneath with some long golden hairs. Antennæ stout, the scape with some long white hairs; the flagellum almost bare. The base of the median segment with its area irregularly striolated; the lateral striations on it straighter and more widely separated; the apex hollowed, smooth, impunctate at base, the apex rough, but without any distinctly defined keels. Pleuræ strongly aciculated, covered with long white hairs; femora clothed with some long white hairs; the tibiæ and tarsi more thickly with golden hairs. Abdomen shining, black, the basal segment reddish, the base with longish white hairs, deeply and widely incised above; the ventral segments like the dorsal. The first recurrent nervure almost interstitial, received immediately in front of the transverse cubital nervure; the second in the apical fourth of the cellule.

b. Green or blue species.

HALICTUS GRANDICEPS, *sp. nov.*

Cupreo-viridis, longe albo-hirsutus; antennis nigris; alis flavo-hyalinis. ♀. Long. 8—9 mm.

Hab. Mussouri (*Rothney*).

Head large, wider than the thorax, coppery green, thickly covered with white pubescence; closely and uniformly punctured; except above, and on the clypeus, where the punctures are more widely separated: the apical half of the clypeus coppery and fringed with golden hairs. Antennæ black; very sparsely covered with microscopic pile. Thorax above closely and rather strongly punctured; the punctures wider apart and larger on the apex of the mesonotum and on the scutellum; the hairs on the mesonotum sparse; on the post-scutellum long and thick; the base of the median segment depressed, finely longitudinally striated; this part in front being bordered by a shining, smooth, glabrous space; the apex has an oblique slope; an elongated deep depression in the middle, into which run two shallow curved furrows from the top, which enclose a triangle at the top. Pleuræ closely and finely punctured; the enclosed space below strongly, the hind wings at top and bottom strongly transversely striolated; the lower part of the metapleuræ at the bottom at the base finely longitudinally striated. The four hinder tibiæ and tarsi thickly covered with long pale fulvous hairs; the femora and the fore legs with the hairs whiter and sparser; the spurs pale fulvous. Wings hyaline but with a distinct fulvous tinge, especially towards the base; the stigma and nervures fulvous, the lower nervure of the costa blackish. Abdomen shining, shagreened, except the base of the basal segment; the apices of the segments fringed with white hairs; the last segment thickly covered with long pale golden hairs; the ventral segments shagreened and covered with long pale hairs at the apex.

HALICTUS ALEXIS, sp. nov.

Viridis, dense fulvo-hirtus; alis hyalinis, stigmate pallide flavo: pedibus longe, dense, pallide pilosis. ♀. Long. 7 mm.

Hab. Barrackpore (*Rothney*).

In its bronzy green coloration it agrees with *H. propinquus*; but it is larger, and the tibiæ and tarsi are not yellowish-fulvous, neither are the tegulæ fulvous.

Head uniformly and closely punctured all over, except on the clypeus where they are fewer and much more widely separated, densely covered with longish pale fulvous hairs; the apex of the clypeus shining, dark bronzy; the labrum covered with long pale golden hairs; the mandibles entirely black; their lower side bearing some long pale golden hairs. Antennæ entirely black; the scape sparsely covered with long pale fulvous hairs; the flagellum almost glabrous. Mesonotum closely and rather strongly punctured, thickly covered with fulvous hairs; the parapsidal furrows distinct; the fulvous hairs at the apex of the scutellum and on the post-scutellum long and thick. The basal curve on the median segment finely and closely rugose; its apex shining and impunctate; the apex of the segment with an oblique slope; shagreened; furrowed down the centre. The propleuræ deeply excavated; the excavation forming an oblique triangle, obscurely striated down the centre and at the apex thickly covered with white hairs; the mesopleuræ strongly and closely punctured; the hairs very long and thick; the metapleuræ finely punctured; the middle finely striated. Legs, especially the hinder, thickly covered with pale fulvous hairs; the femora with the hairs longer and much sparser; the fore femora beneath glabrous, shining, black; the tarsi ferruginous at the apex. Wings clear hyaline, the stigma and nervures yellowish-testaceous, the stigma darker; the second cubital cellule at the top somewhat shorter, at the bottom equal in length to the third at the top, which is there scarcely half the length it is at the bottom; the first recurrent nervure is almost interstitial; the second received near the apical third. Abdomen dark bluish-green; the segments fringed with pale fulvous pubescence; the apical very thickly with longer fulvous hair, except on the furrow in the centre. Ventral segments shining, thickly covered with long pale fulvous hairs.

HALICTUS DISCURSUS, sp. nov.

Long. fere 4 mm.

Hab. Mussouri (*Rothney*).

Comes near to *H. propinquus* Sm., but may be known from it by its smaller size, by the thorax not being thickly covered with white hairs, and by the deep, wide, longitudinal furrow on the base of the mesopleuræ; and by having a short longitudinal furrow at the base of the mesonotum.

Bluish-green, metallic, shining, the knees, tibiæ and tarsi rufo-testaceous; wings clear hyaline, the stigma and nervures pallid, the stigma somewhat darker. The apex of the clypeus is bronzy; the labrum is thickly covered with long golden hairs; the mandibles and trophi testaceous; scape of the antennæ black; the flagellum brownish beneath; covered with white microscopic pile. The area at the base of the median segment finely and closely longitudinally striated except at the apex; which has an oblique slope; the pleuræ shining, impunctate, sparsely covered with white hairs; at the base of the mesopleuræ and above its middle is a wide, deep furrow, extending from the base to the apical third; the metapleuræ finely punctured at the base. Legs thickly covered with white hairs. The second and third cubital cellules are subequal: the first recurrent nervure is interstitial; the second, the third transverse cubital and the cubital nervure from the second transverse cubital, are almost obsolete. Abdomen shining; the base with a wide, deep, longitudinal furrow; the apical segments thickly covered with long white hairs; the fifth segment on either side of the smooth dark testaceous central rima, thickly covered with long pale golden hairs; the ventral segments pale at the apices; the sides of the basal segments and the apices of the apical segments thickly covered with long white hairs; the basal segments in the middle sparsely covered with long white hairs.

c. *Species entirely black; the abdomen banded with white hairs.*

HALICTUS SEPULCHRALIS, *sp. nov.*

Niger; abdomine late albo-balteato; pedibus longe fulvo-hirsutis; alis hyalinis. ♀. Long. 8 mm.

Hab. Mussouri (*Rothney*).

Head black, hardly so wide as the thorax, the face and outer orbits thickly covered with short white pubescence; the vertex and front closely punctured, sparsely covered with long fuscous hairs; the middle and apex of the mandibles piceous. Antennæ black, shining, almost glabrous. Pro- and meso-notum very shining, with only a few microscopic punctures; sparsely covered with short white hairs; the post-scutellum with long white hairs; the median segment at the base with a belt of longitudinal striæ narrowed gradually towards the apex; shining; its apex with an oblique slope, hollowed in the centre, the sides distinctly bordered; the pleuræ and sternum alutaceous, covered with long white hairs. The hind legs covered all over thickly with white hairs; the front four legs less strongly and thickly haired; the femora bare and shining in front; the hairs on the tarsi have a more fulvous tinge; apices of the tarsi rufous. Wings clear hyaline; the nervures fuscous; the stigma paler; the costa darker; the first recurrent nervure almost interstitial; the second received in the apical fourth of the cellule; the tegulæ black, piceous in the middle. Abdomen shining, impunctate; the base of the first segment sparsely covered with longish pale hairs; the second to fifth segments bordered at the apex with white depressed pubescence; the apical segment finely punctured laterally; the rima aciculate, piceous; the ventral segments sparsely covered with long hairs, shining, the basal segments more or less piceous.

HALICTUS PICIPES.

Niger, nitidus; pedibus piceis: capite thoraceque long albohirtis: alis hyalinis; ♂ flagello antennarum subtus bronneo. ♀. Long. 6—7 mm.

Hab. Mussouri (*Rothney*).

Head alutaceous, thickly covered with longish white hairs, which are shorter, sparser, and darker coloured on the vertex. Mandibles piceous before the teeth. Antennæ stout, the flagellum beneath, brownish, darker at the apex. Pro- and meso-notum shining, impunctate, sparsely covered with fuscous hairs; post scutellum thickly covered with white down and with long white hairs. Base of the median segment irregularly longitudinally keeled; those in the middle being wider apart; its apex with an abrupt oblique slope; shining, impunctate; the middle hollowed widely at the top. Pro- and meso-pleuræ closely and finely rugosely punctured; the oblique depression under the fore wings irregularly transversely striolated; the metapleuræ have a blistered appearance, and are covered with long white hairs. Wings clear hyaline; the nervures fuscous; the first recurrent nervure interstitial; the second received in the apical fourth of the cellule. Tegulæ fuscous, lighter coloured round the edges. Legs piceous; thickly covered with longish white hairs; the coxæ black. Abdomen shining, impunctate, the basal segment at the base with some long white hairs; the others fringed with white hairs at the apices; the ventral segments fringed with long white hairs; the apical segment broadly rounded; its sides acutely projecting.

HALICTUS TARDUS, *sp. nov.* (Pl. 4, f. 18).

Niger; longe albo-hirtus: tarsis fulvo-hirsutis: metanoto reticulato: alis hyalinis. ♀. Long. 8 mm.

Hab. Mussouri (*Rothney*).

Head closely and somewhat strongly punctured; densely covered with long soft white hairs; the labrum fringed with long fulvous hairs; the mandibles piceous at the base. Antennæ stout; the scape with longish white hairs; the flagellum with sparse white down. Mesonotum and scutellum shining, bearing widely separated shallow punctures, and, as well as the postscutellum, thickly covered with long white hairs; the median segment at the base closely longitudinally striated, the striæ not all quite straight; the apex of the striated area smooth and impunctate; the apex of the segment strongly reticulated; and with a deep and wide furrow down the centre; the sides sharply and stoutly margined. Propleuræ smooth, shining; the top bluntly triangular; the mesopleuræ rugosely punctured; the metapleuræ at the base alutaceous; the apex reticulated; but with the keels much weaker than they are on the mesonotum.

Legs black; the coxæ and femora sparsely covered with long white hairs; the tibiæ and tarsi much more thickly with fulvous hairs; those of the tarsi having a deeper fulvous tint than those of the tibiæ; the calcaria fulvous. Wings hyaline; the nervures fuscous; the costa and upper part of the stigma black; the lower part of the latter piceous; the second and third cubital cellules at the top and bottom nearly equal in length. Abdomen impunctate, shining, shagreened towards the apex; the basal segment at the base covered with long erect white hairs; the other segments, except the last, fringed with depressed white hairs; the last segment thickly covered with stiff black hair; the apical area fringed with golden hair; the ventral surface thickly covered with long white hairs.

HALICTUS FUNEBRIS, sp. nov.

Long. 9 mm. ♀.

Hab. Mussouri (Rothney).

This species agrees with H. tardus in having the metanotum strongly striolated at the base; and in having the tarsi thickly covered with long golden hairs; but it may be at once known by the apex of the median segment not being reticulated, and having only a few oblique stout keels.

Head shagreened, the clypeus obscurely punctured; covered all over with long white hair; the mandibles piceous towards the apex, bearing beneath a few fulvous hairs. Antennæ stout, bare, the scape with a few long white hairs. Mesonotum coarsely shagreened, opaque; the lateral furrows distinct; covered, as is also the scutellum, with long pale hairs; the scutellum rough at the base, the apex irregularly reticulated; the base of the median segment longitudinally closely and stoutly carinated; the apex with a central keel, from which run a few oblique keels. Legs black, the femora and tibiæ sparsely covered with white hairs; the hairs on the tarsi denser and golden. Wings clear hyaline, the nervures black; the second cubital cellule at the top equal in length to, at bottom shorter than, the third; the first recurrent nervure is received very shortly before the second transverse cubital; the second before the apical third of the cellule. Abdomen shining, impunctate, sparsely covered with fuscous hairs, above and below.

HALICTUS CIRIS, *sp. nov.*

Niger; clypeo, labro, scapo antennarum subtus, tibiis tarsisque, flavis; flagello antennarum subtus brunneo; alis hyalinis. ♀. Long. fere 5 mm.

Hab. Barrackpore (*Rothney*).

Head black, from shortly below the ocelli to the apex of the clypeus thickly covered with short white hairs. Clypeus shining, bearing large widely-separated punctures; at the top, in the middle, the yellow projects triangularly into the black. Mandibles yellow, the apex rufous; the palpi and tongue testaceous. Antennæ stout covered with a pale microscopic pile; the yellow on the under side of the scape with a black mark in the middle. Thorax black, alutaceous, not shining, sparsely covered with a short down; the edge of the pronotum in front and below the tegulæ lemon-yellow; the tegulæ yellow; the base of the median segment longitudinally striolated; the space between the striæ aciculated; the apex of the depression smooth and shining; the apex of the segment oblique; furrowed down the centre; the sides distinctly and strongly keeled. The pleuræ strongly aciculated, opaque, thickly covered with white hairs; the propleuræ excavated. Legs covered with white hairs; the apex of the fore femora and the hinder four knees rufo-testaceous; the tibiæ and tarsi yellow; the hinder tibia rufous before and behind, with a large black line in front. Wings clear hyaline; the nervures pallid yellow; the costa darker; the first and second transverse cubital nervures bullated at the top next to the radial nervure; the first recurrent nervure interstitial; the second received in the apical fourth of the cellule.

HALICTUS VISHNU, sp. nov.

Niger; clypeo, geniculis tarsisque testaceis; alis hyalinis, nervis stigmateque testaceis. ♀. Long. 4 mm.

Hab. Mussouri (Rothney).

Head thickly covered with longish fuscous hairs; the clypeus testaceous, the middle of the mandibles piceous. Antennæ almost glabrous; the flagellum brownish beneath. Mesonotum covered with longish fuscous hairs; closely punctured; the scutellum with the punctures more widely separated; post-scutellum thickly covered with long fulvous hair; the base of median segment shining, glabrous, impunctate, and marked with a few straight longitudinal keels, which do not reach to the bordering carina at the apex of the basal region; the apex has a sharp oblique slope, and is sparsely covered with long hairs. Pleuræ covered with longish white hairs. Legs covered with white hairs; the knees, anterior tibiæ in front and the tarsi testaceous. Wings hyaline, the stigma and nervures testaceous; the first recurrent nervure received in the apical third of the cellule, the second in the second shortly beyond the middle. Abdomen shining, impunctate; the segments above and beneath thickly covered with white hairs; the apical ventral segment with a distinct bordering keel.

HALICTUS SALUTATRIX, sp. nov.

Niger, nitidus; geniculis tarsisque albis; alis hyalinis. ♀ et ♂. Long. 6 mm.

Hab. Mussouri (Rothney).

Head black, closely and minutely punctured; thickly

covered with pale fulvous hairs; the clypeus shining, the punctures, especially towards the apex, more widely separated; the mandibles testaceous in the middle, piceous towards the apex. Antennæ black, covered with a microscopic down. Thorax shining, minutely punctured; covered with white hairs, especially long on the pleuræ, post-scutellum and apex of median segment; the area at the base of the median segment longitudinally striated; the apex at the base very shining; its central furrow not very wide. Propleuræ deeply excavated in the middle and finely striated; the apex at the top, finely striated; the meso- and meta-pleuræ opaque, shagreened, covered with long white hairs. Legs covered with long white hairs, which are especially thick on the hinder four tibiæ and tarsi; the knees of the front pair, the base of the hinder four tibiæ and the tarsi yellowish-testaceous. Wings clear hyaline, the nervures fuscous; the stigma darker at the base; the first recurrent nervure is almost interstitial; the second is received in the apical third of the cellule; the apices of the basal three segments of the abdomen pale, slightly fringed with pale hairs, the apical segments thickly covered with long fulvo-golden hair; the ventral surface, especially towards the apex, bearing long pale fuscous hairs, the apical segment for the greater part aciculated.

HALICTUS BUDDHA, *sp. nov.*

Long. 5 mm. ♀.

Hab. Mussouri (*Rothney*).

Agrees closely with *H. salutatrix*, and, like it, has the base of the tibiæ and the tarsi white; but the front four tibiæ are entirely white; the clypeus broadly white at the apex, the antennæ broadly brownish beneath, the base of the median segment much more strongly striolated.

Head below and between the antennæ thickly covered with white hairs, the front and vertex more sparsely covered with long fuscous hair; the vertex opaque, coarsely shagreened; the apex of the clypeus and labrum yellowish-white; the palpi fuscous; the scape of the antennæ and the second joint entirely black; the others black above, brownish beneath. Pro- and meso-notum shagreened, bearing a short pale down; a narrow longitudinal furrow down the centre of the mesonotum at the side; the scutellum uniformly shagreened like the mesonotum; the post-scutellum rugose, covered with long pale hairs. The curved basal area on the median segment shining, irregularly longitudinally carinate, the keels less distinct in its centre. The pleuræ closely longitudinally striated; the part below the hind wings covered with long white hairs; the propleuræ and the pronotum in front shining, glabrous, impunctate; and sharply separated obliquely perpendicularly behind. Wings clear hyaline, the nervures pale yellowish; the second and third cubital cellules almost equal in length at top and bottom; the first recurrent nervure is almost interstitial; the second is received in the apical third of the cellule. Abdomen shining; the basal segment, except at the apex, covered with long white hairs and without a longitudinal furrow; the segments at the apex fringed with white depressed hairs; the ventral segments bearing long white hairs; the penultimate segment shining and smooth in the middle.

Halictus alphenus, sp. nov.

Nigro-cæruleus; abdomine nigro; femoribus subtus piceis; pedibus longe albo-fumatis. ♂. Long. 6—7 mm.

Hab. Mussouri (*Rothney*).

Head and thorax dark blue; the clypeus black. Head obscurely shagreened, covered with long fuscous hairs; the base of the mandibles piceous. Antennæ entirely black, the flagellum almost bare; the scape bearing long white hairs. Pro- and meso-thorax impunctate, covered with long white hairs; the pleuræ under the wings striated; the base of the median segment obscurely longitudinally striated; its apex oblique, slightly hollowed in the centre, the sides margined. Legs black; the femora with a piceous hue; sparsely haired, the hairs on the tibiæ and tarsi much thicker, especially on the tibiæ which have a fulvo-golden hue; the spurs yellowish-testaceous. Abdomen black, shining, glabrous, except at the apex, which bears fulvous hairs; the ventral segments shining, impunctate, almost glabrous. Wings clear hyaline, the nervures fuscous; the first recurrent nervure almost interstitial; the second received in the apical fourth of the cellule.

HALICTUS GARRULUS, *sp. nov.*

Niger, pallide fulvo-hirtus; geniculis, tibiis tarsisque pallide fulvis; alis fulvo-hyalinis, stigmate fulvo, nervis pallidis. ♂. Long. 7 mm.

Hab. Mussouri (*Rothney*).

Head black; the front and vertex coarsely alutaceous; the lower part of the face densely covered with golden-fulvous pubescence; the mandibles dark piceous, black at the base. Antennæ brownish, the upper part blackish; the scape black, sparsely covered with longish pale hairs. The edge of the pronotum, the mesonotum, fringed all round with pale fulvous hairs; the mesonotum closely and rather strongly punctured; the scutellum with the punctures more widely separated; the post-scutellum thickly covered with pale pubescence, completely hiding its surface; the base of the metanotum strongly longitudinally keeled throughout; the remainder opaque, closely punctured, furrowed in the middle at the apex and thickly covered with long white hairs. Pleuræ opaque coarsely alutaceous; thickly covered with long white hairs. Legs thickly covered with fulvous white hairs; the femora with a piceous tinge beneath; the tibiæ and tarsi fulvous; the tarsi paler. Wings with a faint fulvous tinge; the stigma and costa fulvous; the nervures testaceous; the second transverse cubital nervure largely bullated; the recurrent nervures received shortly beyond the middle. Abdomen shining, impunctate; the segments edged with white pubescence; the ventral surface shining, the first and second segments shining, very sparsely haired; the others much more quickly covered with longish fuscous hairs.

d. Abdomen thickly banded with fulvous hairs.

HALICTUS PULCHRIVENTRIS, *sp. nov.*

Niger: tarsis albis; alis fere hyalinis, apice fumatis. ♂.
Long. 9 mm.

Hab. Mussouri (*Rothney*).

Head black; in front from shortly above the antennæ, densely covered with pale fulvo-aureous pubescence, which completely hides the skin; vertex and front strongly punctured, half shining, sparsely covered with fulvous hairs. Antennæ black, the second and base of the third joint beneath rufous; the scape covered with longish pale fulvous hairs; the flagellum with a microscopic down. Thorax black, a thick band of fulvous pubescence on the pronotum, and a broader belt behind the scutellum of longer fulvous hairs; the mesonotum covered with short fuscous hairs; the median segment from shortly below the basal area covered with long pale fulvous hairs. Mesonotum strongly punctured; a short shallow longitudinal furrow on each side of the mesonotum, originating at the base and reaching to the middle. The basal area of the median segment longitudinally striolated except at the apex, where it has some widely separated punctures on either side of the base of the furrow; the sides of the segment at the area are impunctate; the rest of it with distinctly separated punctures. The mesopleuræ, except at base and apex, thickly covered with long pale fulvous hairs; the metapleuræ covered with a pale fulvous down and more sparsely with long pale fulvous hairs. Wings hyaline, with a slight fuscous tinge; the apex from the apex of the radial cellule smoky; the nervures fuscous; the stigma darker at the top; both the recurrent nervures are received about the same relative distance beyond the middle. Legs thickly covered with longish fulvous hairs, the tarsi and the anterior tibiæ in front testaceous. Abdomen punctured, strongly towards the base, more weakly towards the apex; the segments shining at the base, sparsely covered with long fuscous hair; the fourth segment thickly covered with depressed fulvous pubescence; the fifth obliquely depressed; covered with long fulvous hair; the last segment ferruginous.

HALICTUS TAPROBANÆ, *sp. nov.*

Long. 6 mm.

Hab. Ceylon (*Rothney*).

Agrees very closely in form and coloration with *H. pulchriventris*, but is much smaller; and otherwise easily separated by the striated area of the median segment being distinctly and triangularly produced at the apex, which is not the case in *H. pulchriventris*; it is also not hollowed at the apex.

Antennæ black, the flagellum brownish on the underside, covered with a pale microscopic down. The face below the antennæ and the sides as high as the lower ocellus, thickly covered with pale fulvo-aureous pubescence; the mandibles broadly ferruginous in the middle. Thorax black; the mesonotum and scutellum closely punctured, the edge of the pronotum behind and a belt behind the scutellum thickly covered with long pale fulvous pubescence; the mesonotum covered with short, the scutellum with long, dark fulvous hairs. The area on the median segment triangularly produced in the middle, and longitudinally striolated; the central striæ not reaching to the apex; the area bare, the rest of the segment covered with long pale fuscous hairs. Mesopleuræ thickly covered with fulvous hairs; the hairs on the metapleuræ sparser and longer. The front femora entirely, beneath and above, and the four posterior entirely on the under side, obscure testaceous, as are also the apices of the tibiæ; the basal joint of the tarsi white; the others white, but with a testaceous tinge. Wings hyaline, the apex slightly infuscated, the stigma and nervures testaceous, the former black at the extreme base. The first recurrent nervure is received shortly before, the second shortly beyond, the middle of the cellule. Abdomen shining, the segments shagreened; their apices thickly fringed with pale hairs; the last segment testaceous. Ventral segments shining, sparsely covered with long hairs, their apices white; the last testaceous; the penultimate with a shallow depression, wide at base, becoming gradually narrower towards the apex.

ANDRENA ROTHNEYI, *sp. nov.*

Nigra; abdominis basi late rufo; capite thoraceque rufis; alis hyalinis. Long. 10—11 mm.

Hab. Mussouri (*Rothney*).

Antennæ entirely black: the scape sparsely covered with long fuscous hairs; the flagellum with an obscure microscopic pile. Head covered with long pale fulvous hairs, more sparsely on the centre of the clypeus and of the vertex; the inner orbits of the eyes bordered with a band of pale fulvous depressed pubescence; the apex of the clypeus fringed with golden hairs: the clypeus sparsely; the inner orbits more closely punctured; the vertex finely and closely longitudinally striated from the hinder ocelli, behind which it is smoother and shining; the occiput thickly covered with long pale golden hairs. The mandibles are piceous before the middle; the joints of the palpi are white at the base. Thorax bearing pubescence of moderate length; fulvous above: that on the pleuræ paler; finely punctured; a longitudinal furrow on either side of the mesonotum, a very indistinct one down the middle of the scutellum, which at the apex is fringed with long fulvous hairs. The middle of the median segment bearing a large somewhat triangular opaque shagreened space not uniting with the apex; the sides shining and smoother, and bearing long fulvous hairs; the apex being similarly clothed. The pleuræ shagreened, somewhat shining; the apex of the pro- and of the meta-pleuræ thickly covered with long pale fulvous hairs; the sternum sparsely covered with long pale fulvous hairs. Legs black; the hairs pale fulvous. Abdomen with the basal two segments ferruginous above and beneath; shining, impunctate, almost glabrous; the other segments black, their apices fringed with pale fulvous hairs: the hypopygium aciculated, the sides sharply bordered; the centre with an elongated raised space, sharply pointed towards the apex; the ventral segments covered, but not very thickly, with long fulvous hairs, which are longest towards the apex. Wings with a faint fuscous tinge; the nervures black.

Andrena communis Sm. ("North India, Masuri; taken at an elevation of 7,000 ft."), resembles this species in coloration; but it has the antennæ fuscous beneath: the apical margin only of the first abdominal segment is ferruginous, which colour also extends to the third. Both, as also the following species, resemble the European *Andrena cetii*. Smith points out this resemblance in regard to his *A. communis* (Descr. New Sp. Hym. p. 51).

ANDRENA MALIGNA, *sp. nov.*
Long. 9 mm. ♂.
Hab. Mussouri (*Rothney*).

Agrees with *A. Rothneyi* generally in coloration; but the basal abdominal segment is only ferruginous at the apex, while the third is entirely ferruginous; the hairs on the head and thorax much thicker and longer and uniformly distributed; the pleuræ much more strongly punctured, the mesopleuræ also being obliquely striolated behind. Judging from the description it can hardly be the ♂ of *A. communis*.

Antennæ entirely black, the scape sparsely covered with long fuscous hairs; the flagellum opaque, almost glabrous. Head large, distinctly wider than the thorax; black; thickly covered with long fulvous hairs; the vertex except at the sides and behind, closely, somewhat obliquely, striated; the clypeus with large, clearly separated, punctures; tips of mandibles piceous. Thorax densely covered all over with long fulvous hairs; propleuræ deeply excavated; the mesopleuræ opaque; the base obscurely punctured, the apex obliquely striated; the base of the metapleuræ with a deep shining, oblique depression. Legs densely covered with long pale fulvous hairs. Wings hyaline, but with a distinct fuscous tinge; the first and second transverse cubital nervures are distinctly bullated at the base and apex and roundly curved, the third on the lower side. Abdomen thickly covered with longish fulvous hairs; the first segment black, except at the apex; the second ferruginous, except a black stripe before the apex; the third is entirely ferruginous; the apical three segments thickly covered with long fuscous hairs; the basal five ventral segments ferruginous; the two basal segments with a black line in the centre which bifurcates on the third to fifth; the sixth and seventh black; the seventh thickly covered with long fulvous hairs.

ANDRENA RETICULATA, *sp. nov.*

Nigra; capite thoraceque pallide fusco-hirsutis; metathorace reticulato; alis hyalinis, nervis fuscis. ♀. Long. 12 mm.

Hab. Mussouri (*Rothney*).

Head black; covered with long cinereous hairs, except on the clypeus; and having them sparser on the front; the clypeus strongly punctured all over; the extreme apex transverse, shining, impunctate; immediately below the antennæ is a clearly defined space, a little longer than broad, the apex transverse, the sides straight, smooth, and shining. Mandibles deeply grooved; the tips piceous. Antennæ black, covered at the apex with a pale down. Thorax covered all over with pale fulvous hairs, which are paler and longer and thicker on the pleuræ and sternum. Mesonotum strongly punctured, the punctures in the middle more widely separated than on the sides; the scutellum punctured; the punctures smaller than on the mesonotum, and very sparse in the middle at the base; the base of the median segment with stout distinctly separated longitudinal striæ, those in the middle being more widely separated than those at the sides; the apex on either side of the central hollow, strongly transversely striated. Pleuræ with the punctures somewhat less in size than they are on the mesonotum. Femora covered with longish pale hair; the hair on the tibiæ and tarsi shorter and thicker, the spurs rufo-testaceous. Wings hyaline, the nervures blackish. Abdomen shining, impunctate; the segments narrowly lined with silvery hairs.

ANDRENA SÆVISSIMA, *sp. nov.*

Long. 12 mm. ♂.

Hab. Mussouri (*Rothney*).

A larger and stouter species than *A. phædra*, which agrees with it in the structure of the median segment; but the present species differs from it in being larger and stouter; the base of the median segment is more strongly reticulated, the central fovea larger and deeper; and the apex is strongly transversely striolated.

Head black; the front and vertex sparsely covered with long pale fulvous hairs; the clypeus almost bare; the vertex closely punctured; the clypeus with the punctures larger and more distinctly punctured below; projecting between the antennæ is a raised space, very smooth, shining, and triangular at the top; transverse at the bottom and with a few large punctures. Antennæ black, almost glabrous. Thorax densely covered with long pale fulvous hairs; the median segment at the base almost glabrous; the triangular space in the middle at the base aciculate, the apex shining, impunctate; the sides more strongly aciculated. Pleuræ shining, impunctate. The hairs on the femora fulvous; on the tibiæ and tarsi blackish: the middle tarsal joints testaceous. The wings have a decided fusco-violaceous tinge beyond the transverse basal nervure; the nervures and stigma in the centre fuscous; the tegulæ pale testaceous. The basal segment of the abdomen thickly covered with long pale testaceous hairs; the other segments thickly covered with black hairs, except at the apices which are fringed with white depressed hairs; the basal half of the ventral segments smooth, glabrous; the apical fringed with longish black hairs; the pygidium glabrous, shining, impunctate; the lateral furrows wide, covered all over with long pale fulvous hairs, closely and rather strongly punctured; the punctures larger and more widely separated towards the apex; the base of the scutellum almost impunctate; the base of the median segment with short stout keels; those at the sides being longer and sharply bent in the middle; the middle before the basal keels coarsely reticulated; in front of this again is a large, somewhat pear-shaped, shining, impunctate depression;

the rest of the segment is rugosely transversely punctured. The propleuræ have a dense curve of pale fulvous hair; the propleuræ below and at the junction with the mesopleuræ are widely furrowed, the furrow being transversely keeled. Legs thickly covered with pale hairs, which are darker on the hind legs. Wings clear hyaline; the nervures fuscous; the second cubital cellule at the bottom as long as the third; the first recurrent nervure is received very shortly beyond the middle; the second in the apical third of the cellule. Abdomen shining; the segments above and beneath lined with white depressed hair; the basal segment with a deep, wide, longitudinal furrow.

The present species agrees very closely with *A. reticulata*, but it may be known from it by the deep furrow on the basal segment of the abdomen, by the large shining, deep depression at the middle of the median segment; which in *A. reticulata* is smooth, shining, and without any depresion.

ANDRENA MEPHISTOPHELICA, *sp. nov.*

Long. 11—12 mm.

Hab. Mussouri (*Rothney*).

Is related to *A. sævissima*; but may be known from it by its smaller size; by the much stronger and closer punctation of the clypeus; by the hairs on the median segment being much longer, closer and fulvous in colour, differing from the hairs on the mesonotum, which are much paler.

Head, except on the clypeus, covered thickly with long hairs, pale beneath, darker on the front and vertex. Clypeus with large punctures, which are much sparser on the apex, its apex being almost clear of them and fringed with dark fulvous hairs; the base of the mandibles aciculate. The antennæ almost bare; the flagellum from the second joint brownish beneath. Pro- and meso-thorax thickly covered with long pale fulvous hairs, the scutellum with only long hairs behind, and almost without punctures; the median segment, except a triangular space in the middle at the base, thickly covered with rufo-fulvous hairs, which completely hide the texture; the triangular bare space at the base opaque, rugosely aciculated; with an indistinct keel down its centre. The upper part of the propleuræ aciculated. Legs thickly covered with pale fulvous hairs; the calcaria white. Wings clear hyaline; the nervures dark fuscous; the first recurrent nervure is received shortly beyond the middle of the cellule; the second about the same distance from the third transverse cubital nervure. The first and second dorsal segments of the abdomen are covered with long pale fulvous hairs; the others have the hairs darker and shorter, and the second, third, and fourth are fringed at the apex with glistening white hairs; the ventral segments are broadly fringed at the apex with long pale hairs.

ANDRENA GRACILLIMA, *sp. nov.* (Pl. 4, f. 19).

Nigra: capite thorace abdominisque basi longe fulvo-hirtis; alis fumatis, basi fere hyalinis. ♀. Long. 15 mm.

Hab. Mussouri (*Rothney*).

Head deep black; the occiput thickly covered with long fulvous hairs; the front with hairs almost as long, but somewhat shorter. The front from the hinder ocelli closely longitudinally striated; a sharp keel runs down from the ocelli; the clypeus shining, the punctures close at the base, becoming more widely separated towards the apex; which is in the middle almost free from them; the apex slightly projecting. Antennæ black; the flagellum almost glabrous; the fifth and following joints brownish beneath. Pro- and meso-thorax and the median segment punctured; the scutellum with the punctures more widely separated, especially in the middle at the base; the median segment has an oblique slope; the extreme base has short stout longitudinal keels; below this it is reticulated; the rest strongly transversely striolated. Pleuræ rather strongly punctured, covered with long pale hairs; the hairs on the metapleuræ longer and thicker. Legs black; thickly covered with pale fulvous hairs; those on the hind legs thicker and longer. Wings hyaline, the nervures dark fuscous; the first recurrent nervure received very shortly beyond the middle of the cellule. Abdomen shining; the segments fringed with white pubescence, above and beneath; the ventral segments more strongly punctured than the dorsal.

ANDRENA MOROSA, *sp. nov.*

Nigra; capite thoraceque longe pallide hirtis; abdominis basi ferrugineo-maculato; alis hyalinis. Long. 12—13 mm.

Hab. Mussouri (*Rothney*).

Head, except the clypeus, thickly covered with long greyish hair; opaque, and coarsely alutaceous, the front closely longitudinally striated; the clypeus shining, almost glabrous, and bearing distinctly separated punctures; the labrum broadly and roundly incised at the apex. Thorax thickly covered with long greyish hairs, more sparsely on the mesonotum; the mesonotum and scutellum shining, almost impunctate. Median segment coarsely alutaceous; the base rugosely longitudinally striolated; the centre with a shallow longitudinal furrow; the sides bare, and apex thickly covered with long grey hairs. Pleuræ alutaceous, covered with long grey hairs. Legs, especially the hinder pair, thickly covered with long greyish hairs; those on the hind legs being darker; the calcaria white. Wings hyaline, but with a slight fuscous tinge, especially in front; the nervures fuscous. Abdomen shining, impunctate; the first and second dorsal segments at the apex piceous; the second to fourth segments at the apices fringed with pale hairs; the fifth segment with the fringe thicker and longer and dark fulvous; the sixth segment similarly clothed at the sides; the hypopygium alutaceous; the centre triangularly raised, but not sharply. The second ventral segment rufous, black in the centre, the black mark being dilated at the apex; the third and fourth black, piceous at the base.

ANDRENA PHÆDRA, sp. nov.

Long. 8 mm. ♂.

Hab. Mussouri (*Rothney*).

This species is very closely allied to *A. reticulata*, and has, like it, the base of the median segment reticulated; but its apex is not transversely striolated; while in its centre, below the reticulated part, is a deep, shining, impunctate, and somewhat triangular space; the widest part of which is at the base.

Head densely covered with long greyish hairs; closely rugosely punctured; the clypeus strongly punctured, with the punctures more widely separated; depressed at the apex and with very few punctures; the mandibles deeply grooved; their teeth piceous. Antennæ black, towards the apex with a fuscous down. Thorax densely covered with long pale fulvous hairs; the mesonotum and scutellum bearing large, clearly-separated punctures except the former in the middle at the apex and the latter at the base; the base of the median segment with a band of short longitudinal keels; and a second band of similar keels in the middle behind the first; and from the centre of this runs a short deep, shining, somewhat triangular depression; the rest of the segment rugosely punctured. Mesopleuræ with large punctures; a curved furrow above the middle; above which is a large, smooth, impunctate space; the part immediately below the wings being coarsely punctured; the metapleuræ closely rugose.

Legs covered with long white soft hairs; especially thick and close on the tarsi. Wings clear hyaline; the nervures dark fuscous; the second and third cubital cellules at the bottom almost equal in length; the first recurrent nervure is received shortly before the middle; the second in the apical third. Abdomen with the apices of the segments fringed with white hairs; the basal segment sparsely covered with long white hairs; at its base is a deep triangular depression; the ventral segments shining; their apices fringed with white hair.

ANDRENA SODALIS, *sp. nov.*

Long. 8—9 mm.

Hab. Mussouri (*Rothney*).

Agrees with *A. phædra* and *A. reticulata* in having the median segment reticulated at the base; from the former it may be known by there being a triangular keel and spot in the middle of the segment at the base, in which there is no deep triangular depression; from the latter by its smaller size and by the median segment not being transversely striated.

Head densely covered with long pale fulvous hairs, hiding the sculpture; the front and vertex finely and closely rugose; the clypeus strongly punctured, the punctures distinctly separated; the mandibles deeply grooved; their teeth piceous. Antennæ black; the apex with a microscopic down. Thorax above thickly covered with fulvous hairs; the sides and head with longish pale hairs; the mesonotum closely and rather strongly punctured, less closely in the middle towards the apex; the scutellum punctured pretty much as the base of the mesonotum, and covered with longer hairs. Median segment with an oblique slope; the base with a double row of short thick keels; below these is a large wide triangular shining reticulated space; the rest of the segment rugosely punctured. Legs thickly covered with white hairs; the spurs white. Wings clear hyaline; the tegulæ sordid testaceous; the nervures dark fuscous; the second and third cubital cellules at the bottom subequal; the first recurrent nervure is received shortly beyond the middle; the second in the apical third. Abdomen black; the basal segment sparsely covered with long pale hairs; the others belted with white depressed hair at the apex; the ventral segments fringed with white hair; the apical bordered with piceous.

ANDRENA ANONYMA, *sp. nov.*

Long. 11—12 mm.

Hab. Mussouri (*Rothney*).

Head black; the front and vertex covered with long pale hairs, the latter alutaceous, except at the top of the eyes, where it is smooth and shining; in front of this smooth space is a spot of dark fulvous hairs; the clypeus strongly punctured, especially towards the base, the apex broadly shining, smooth, with some widely separated punctures, the centre almost impunctate. Antennæ entirely black; the flagellum almost glabrous. Thorax covered with long fulvous hairs, which are paler on the pleuræ; the mesonotum and scutellum almost impunctate, shining; the median segment alutaceous, with a gradually rounded slope; at the base is a shallow indistinct longitudinal furrow. Legs thickly covered with pale hairs. Wings clear hyaline, the stigma and nervures fuscous: the first recurrent nervure is received in the middle, the second in the apical fourth of the cellule. Abdomen shining, smooth, impunctate, the segments fringed with white hairs at their apices; the apical segments thickly covered with fuscous to dirty white hairs; the ventral segments fringed with long white hair.

APIDÆ.

NOMADA CEYLONICA, *sp. nov.*

Ferruginea; capite thoraceque late nigro-maculatis; abdomine flavo-bimaculato; alis fuscis, basi fere hyalinis. ♀. Long. fere 6 mm.

Hab. Ceylon (*Rothney*).

Head black; the orbits narrowly, the clypeus, mandibles except at the apex, ferruginous; coarsely punctured; the front and vertex covered with long fuscous, the face more thickly with shorter, white hairs; the apex of the clypeus shining, impunctate. Antennæ rufous; almost bare; the flagellum blackish above; the front projecting sharply between the antennæ. Thorax ferruginous, coarsely punctured, rather thickly covered with white hairs; a broad central and two narrower black continuous bands on the mesonotum; the metanotum entirely, the propleuræ, except at the top, the mesopleuræ below the tegulæ, under the wings, and at the apex (but the latter with a long ferruginous mark at the top), the metapleuræ and the sternum, black. The curved furrow in front of the middle coxæ is deep, and the part enclosed by it is much less strongly punctured than the rest of the sternum. The scutellum is strongly punctured and longitudinally depressed down the middle; the post-scutellum is of a paler colour. The median segment is entirely black; the basal area almost rugose; the sides at the front of it very thickly covered with long white hairs. Legs rufous, covered with white hairs; the greater part of the hinder coxæ, the base of the hinder femora, above and beneath, and the hind tarsi, black. Wings fuscous, paler at the base; the stigma fuscous, lighter in the centre. Abdomen shining, impunctate; black. The first segment with a dull ferruginous band before the apex; the second segment dull ferruginous, black in the centre and with a large yellow mark at the side; the ventral surface ferruginous, marked with black.

A form of what is no doubt the same species has only the central line on the mesonotum black; the median segment broadly black only down the middle; the pleuræ and sternum without black; the abdomen above almost entirely black, except the yellow marks, and the hinder femora without black.

ANTHIDIUM FLAVIVENTRE, *sp. nov.*

Flavum, nigro-maculatum; vertice nigro, flavo-maculato; pedibus flavis; alis hyalinis. Long. 5 mm.

Hab. Poona (*Wroughton*).

Head yellow, the vertex from the antennæ to shortly behind the eyes, the black surrounding them entirely narrowly behind; on the vertex is a large yellow mark, broader than long, in the centre between the antennæ and the ocelli; strongly punctured, sparsely covered with white pubescence. Mandibles yellow, the teeth black. Antennæ black, shining, the flagellum obscure brownish beneath. Thorax black, strongly punctured; a large mark in front of the tegulæ; on each side of the mesonotum at the base is a thick straight line which curves round the tegulæ to their end; on each side of the median segment is a large yellow mark, obliquely truncated at the apex, leaving a somewhat triangular black mark in the middle at the base, the apex of the median segment transverse, the sides oblique. Pleuræ coarsely punctured; behind covered with white hairs. Legs yellow, thickly covered with white hairs, the hinder femora broadly black at the base. Wings infuscated at the apex; the nervures black. Abdomen above black, coarsely punctured; on the basal five segments are broad yellow lines, which become gradually broader until, on the sixth, they almost unite; ventral surface lemon-yellow, rugose, thickly covered with short white hairs.

STELIS PARVULA, sp. nov.

Nigra; dense albo-hirsuta; tegulis abdomineque albo-maculatis; alis hyalinis. Long. 4 mm.

Hab. Barrackpore (*Rothney*).

Head thickly covered with longish white hairs; the rest of the head covered with similar hair; but not so thickly; the tips of the mandibles piceous; punctured. Antennæ with the scape covered closely with moderately long pubescence; the flagellum with a microscopic pile. Pronotum finely, the mesonotum coarsely, punctured; the pronotum in front fringed with long white hair; the mesonotum in front is also fringed with long white hair; the rest of it has the pubescence sparser and shorter; the scutellum nearly as strongly punctured as the mesonotum; its apex entire, rounded; its sides broadly white. Mesopleuræ thickly covered with white hair; the propleuræ slightly pilose; the meta- as thickly haired as the meso-pleuræ; the base of the median segment thickly covered with long white hairs; its apex hardly pilose. Legs black; thickly covered with long white hairs, the knees and apices of the tarsi rufous; the calcaria yellowish-white. Wings clear hyaline; the nervures fuscous; the stigma darker; tegulæ large, yellow, a large black mark in the centre. Abdomen thickly covered with white hairs, especially towards the apex; the sides of the segments with longish, moderately broad yellow marks; the basal two segments narrowly lined with yellow; the third to fifth segments bear two elongated yellow marks; the apical segment has an elongated mark at the sides, and two somewhat roundish ones in the centre. Ventral segments thickly covered, especially at the apices, with long white hairs; their sides lined with yellow.

COELIOXYS.

The species of this genus known to me from India may be separated as follows:—

1 (6) Thorax coriaceous, the punctures not distinctly separated.
2 (5) With metanotal spines.
3 (4) The metanotal spines long, sharp, curved.
basalis
4 (3) The spines short, blunt, straight, wings subhyaline.
apicalis
5 (2) Without metanotal spines.   *argentifrons*
6 (1) Thorax coarsely punctured, the punctures distinctly separated.
7 (8) Thorax with six marks of white pubescence, the scutellum much more finely and closely punctured than the mesonotum.   *sexmaculata*
8 (7) Thorax not maculate, the scutellum not more coarsely punctured than the mesonotum.
9 (10) Apex of scutellum projecting in the middle.
fuscipennis
10 (9) Apex of scutellum almost transverse, not projecting in the middle.   *confuscus, cuneatus*

COELIOXYS SEXMACULATA, sp. nov.

Long. 11 mm. ♀.

Hab. Barrackpore (*Rothney*).

Head in front densely covered with white pubescence, which is thicker at the sides; the orbits behind except at the top with similarly coloured hairs; the vertex and front strongly punctured. Antennæ black, almost glabrous. Pronotum strongly punctured, lined with white pubescence; mesonotum more coarsely punctured; the scutellum more closely and finely punctured; there are two white spots on the base of the mesonotum, two on the base of the scutellum, and a smaller one behind the tegulæ. Scutellar spines stout. Mesopleuræ and metapleuræ thickly covered with white pubescence; strongly punctured. Wings fuscous; more lightly coloured at the base. Legs black; the tarsi beneath thickly covered with golden pubescence; the spurs black. The basal segment closely and rather strongly punctured, margined with silvery white pubescence; the transverse furrow on the second and third segments rugose; the apical segment above closely punctured, keeled down the centre; the keel indistinct at the base, becoming thicker towards the apex, where it is depressed on either side of it. The ventral segments punctured; a band of white pubescence down the centre of the basal; the others transversely banded with silvery pubescence.

Comes near to *C. fuscipennis*, but that species wants the white marks on the mesonotum, which has also the punctures more distinctly separated, this being especially noticeable on the scutellum, where they are round and deep, and not, or hardly, touching each other, whereas in *C. sexmaculata* they are much coarser and closer, forming a rugose surface.

ANTHOPHORA DEIOPEA, *sp. nov.*

Nigra, longe dense pallide hirta; capite nigro & facie alba. Long. 13 mm. ♀.

Hab. Mussouri (*Rothney*).

Head black, thickly covered with long pale grey hairs, especially on the front and vertex; the labrum fringed with golden hairs; the mandibles ferruginous, black at the apex. The vertex behind the front ocellus bare, shining, broadly depressed. Thorax thickly covered with long grey hairs all over. Legs: the femora and tibiæ dark rufous; the former sparsely covered with long white hairs, the front four tibiæ covered densely behind with pale fulvous hairs; the hairs on the hinder tibiæ much longer, thicker, and of a brighter fulvous tint; the tarsi rufous, thickly covered with long golden hairs at the base. Wings hyaline, with a faint fuscous tinge: the costa and nervures blackish; the first recurrent nervure is received shortly before the second transverse cubital nervure; the second is interstitial. Abdomen above and at the sides thickly covered with long pale fulvous hairs; the penultimate segment rufous at the apex; the apical ferruginous, black at the apex, the base closely transversely striated; the sides, especially towards the apex, broadly furrowed; abdominal segments black, the base and apex broadly ferruginous; the segments at the apices thickly fringed with fulvous hairs.

The ♂ is covered all over with long hoary hairs; the clypeus, except at the sides and the inner orbits, cream-yellow; the extreme apex piceous, the mandibles cream coloured: the tips black, ferruginous in front of the black; the labrum black, covered with white hairs; the ventral segments are coloured as in the ♀; this being also the case with the legs, which bear long white hairs.

MEGACHILE SAMSON, sp. nov.

Nigra: thorace abdominisque basi rufo-hirsutis; alis fusco-violaceis. ♂. Long. 25 mm.

Hab. Himalayas.

Head deep velvety black, opaque, coarsely alutaceous, thickly covered with black hairs, which are longest and thickest on the front and at the base of the clypeus, which is short, coloured and haired like the vertex, and projecting in the middle into a stout, large, somewhat triangular thickly-haired tooth; its apex shining, and smooth at the sides; the labrum large, as long as the space between the ocelli and the apex of the clypeus, covered with a dull golden down and with some long black hairs; its apex bearing much longer hairs. Mandibles very large, opaque, the middle above with some elongated punctures and elongated striæ; the apical tooth large; the basal rounded in the middle. Antennæ black, glabrous. Thorax opaque, closely rugosely punctured; above thickly covered with rufous hair, this being also the case on the upper part of the pleuræ; the hairs on the lower part are much darker; on the sternum fulvous, the latter broadly depressed in the middle at the base. Wings smoky, darker and more violaceous at the apex; the base with a slight yellowish tinge; the costa, stigma, and nervures black; the last with a yellowish tinge in the middle of the wing; the recurrent nervures are both received at the same distance from the transverse cubitals. Legs thickly covered with stiff black hairs; the anterior four tibiæ end above in a large stout somewhat triangular process, which ends in a small curved point. The basal abdominal segment broadly depressed in the middle above; the sides, base, and apex thickly covered with rufous hairs; the second segment depressed at the base, fringed with fulvous hairs, this being also the case with the third at the sides; the second and following segments thickly covered with stiff black hairs. Ventral surface thickly covered with long stiff black hairs. The hinder calcaria are short and thick.

This *Megachile* is, next to *M. Pluto* Sm. (from Bachian), the largest of the species from the Oriental Region. Smith's species is 18 lines in length, that being however the length of a ♀, the only sex known to its describer (*Trans. Linn. Soc.*, V., *1860, 133*). Our species is also apparently related to *M. monticola* Sm., but I cannot make it agree with Smith's description.

Megachile Hornei, sp. nov.

Nigra; facie longe fulvo-hirta; pedibus anticis rufo-testaceis, alis fumatis. ♂. Long. 17 mm.

Hab. Mussouri (*Rothney*).

Head large; below the antennæ and the orbits to the ocelli, densely covered with long fulvous hairs. Mandibles black, covered with long fulvous hairs, and with large clearly separated deep punctures. Antennæ black; the scape closely punctured. Thorax black, closely and rather strongly punctured; the pronotum and prosternum thickly covered with long fulvous hairs; the rest of the thorax thickly covered with black hairs, that on the mesonotum shorter, on the median segment as long as the fulvous hairs on the pronotum. On the base of the median segment is a dull impunctate area dilated into a sharp point in the middle; the rest alutaceous, obscurely punctured; the pleuræ opaque, obscurely punctured, thickly covered with long black hair. Legs: the anterior femora and tibiæ, the base of the middle femora and the lower part of the middle tibiæ in front, fulvo-testaceous; the anterior four tarsi thickly covered with pale fulvous hairs; the hinder tarsi covered with longer, thicker, and whiter hair. Wings at the apex fuscous, with a faint violaceous tinge, lighter below the stigma and at the base. Abdomen black; above closely punctured, thickly covered with black hairs, very long at the base and the apex; the base semicircularly incised; its apex very smooth, shining, and with a bluish tinge; a narrow furrow extends from the middle to the base; the apical dorsal segment deeply incised in the middle; the apex flat; triangularly incised in the middle. The ventral segments fringed with short fulvous hairs at the apex; the apical segment depressed, coarsely punctured, covered at the base with long fulvous hair.

Allied to *M. anthracina* Sm.=*M. fasciculata* Sm., ♂ cf., Horne and Smith (*Trans. Zool. Soc. vii.* (*1872*), *p. 179*). The fore coxæ are sharply triangularly produced in front.

MEGACHILE SYCOPHANTA, *sp. nov.*

Nigra; capite thoraceque longe cinereo-hirtis; abdomine subtus longe albo-hirto; alis hyalinis. ♀. Long. 13—14mm.

Hab. Mussouri (*Rothney*).

Head thickly covered with long hairs; fulvous above, cinereous below the antennæ. Clypeus coarsely punctured; a shining, impunctate longitudinal line down the middle; the teeth on the mandibles blunt, rounded; the inner side of the mandibles smooth; the rest irregularly striated and punctured; the lower side bearing long golden hair. Antennæ black, shining, smooth. Thorax closely and rather strongly punctured; the mesonotum (but very sparsely in the middle) with long pale fulvous hair; the scutellum almost impunctate, the post-scutellum fringed with long pale fulvous hairs; the pleuræ and sternum punctured like the mesonotum and covered with long white hairs. Legs black; the hairs of the tibiæ and tarsi on the inner side thick, deep golden; on the rest of the legs the hairs are longer, sparser, and pale silvery; the calcaria pale. Wings hyaline, the nervures blackish. Abdomen closely punctured, the basal segments thickly covered with long pale fuscous hairs; the third and following segments fringed with short silvery hairs, the third and fourth deeply depressed; the last segment coarsely and closely rugose; the ventral scopa apparently white, but the colour hidden by reddish pollen.

The ♂ is similarly coloured to the ♀; there is no spine before the front coxæ; the apical abdominal segment has no spines; it is broadly and roundly incised; this being also the case with the fourth, and, to a less extent, with the third segment; above the apical segment is depressed at the apex.

MEGACHILE IMPLICATOR, sp. nov.

Nigra, longe fulvo-hirta; alis hyalinis, apice fumatis. ♂. Long. 9 mm.

Hab. Mussouri (*Rothney*).

Antennæ black, the scape with fulvous hairs; the flagellum almost bare. Head densely covered all over with long fulvous hairs, being especially thick and long on the face and vertex. Mandibles entirely black, coarsely punctured at the base, where they have a few fulvous hairs. Thorax thickly covered all over with long fulvous hairs; the mesonotum strongly and closely punctured. Legs thickly covered all over with long pale hairs; those on the under side of the hinder tarsi inclining to fulvous; the claws piceous. The wings are almost hyaline to the end of the radial cellule, when they become fuscous; the tegulæ black. Abdomen black; the basal segment at the apex thickly; the second and third more narrowly fringed with long fulvous hairs; the apical segments with long black hairs; the ventral segments fringed at the apices with long white hairs; the apical broadly and roundly incised.

The anterior four coxæ and the base of the femora are strongly punctured; the apex of the latter very smooth and shining and with a piceous tinge.

Comes near to *M. lanata*; but, apart from the difference in coloration, the latter may be known from it by the strongly rugosely punctured pleuræ and sternum.

MEGACHILE ALBOLINEATA, sp. nov.

Nigra; abdomine albo-lineato; femoribus posterioribus rufis: alis hyalinis. ♀. Long. fere 10 mm.

Hab. Ceylon (*Rothney*).

Head rather closely punctured; the inner and outer orbits of the eyes broadly covered with white hairs; the clypeus projecting; smooth, shining, impunctate, and slightly notched at the apex; mandibles coarsely punctured, their lower edges smooth, shining, and impunctate, and fringed with long golden hairs. Mesonotum strongly punctured, thickly covered with long fuscous pubescence; the sides, base, and apex of the tegulæ thickly with long white hairs. Base of median segment shining, impunctate; the rest of it closely punctured, and covered with long white hairs. Femora on the lower side sparsely covered with soft white hairs; the tibiæ more thickly; the tarsi still more thickly covered with fulvous pubescence, especially the hinder four. Wings clear hyaline, the nervures and stigma black. Abdomen above rather strongly and closely punctured, black; the segments at the apex densely fringed with white pubescence; the transverse furrows on the second and third segments deep, shining, impunctate. Ventral fringe long; clear white.

MEGACHILE MALIGNA, sp. nov.

Nigra; femoribus rufis; abdominis scopa fulva; alis hyalinis. ♀ et ♂. Long. 8—9 mm.

Hab. Mussouri (Rothney).

Head black, the front and face thickly covered with long pale fulvous hair; the vertex closely punctured, the hairs much sparser and of a deeper fulvous tint. Mandibles strongly punctured, thickly covered at the base with long pale fulvous hairs; the apical two teeth stout; the part at their base piceous. Antennæ entirely black. Thorax coarsely alutaceous, thickly covered with white hairs, which are especially long on the sides and metanotum, where they are of a paler tint. Median segment alutaceous, its slope rather abrupt. The fore femora black; the lower side entirely and the upper side at the base above, and the hinder four pairs, rufous; the tibiæ and tarsi thickly covered with white hairs, which have a fulvous tint on the metatarsus; the spurs pale testaceous; the apex of the hinder tibiæ piceous on the outer side. Abdomen black; above closely punctured; the segments fringed with silvery pubescence; the scopa pale fulvous; the basal ventral segment rufous.

The ♂ is similarly coloured to the ♀; except that the ventral segments are rufous, except the apical one, which is depressed broadly in the middle, and cleft slightly and roundly; on either side are three stout teeth, which become successively, but not much, shorter. On the sternum in front of the fore coxæ are two stout projecting plates, curved on the inner side, straight and slightly oblique on the outer.

MEGACHILE PULCHRIPES, sp. nov.

Nigra, longe argenteo-pilosa; coxis, trochanteribus, femoribus tibiisque posticis, rufis; alis hyalinis, stigmate fusco. ♀. Long. 6 mm.

Hab. Mussouri (*Rothney*).

Head coarsely punctured; the inner orbits thickly covered with white depressed pubescence; the rest of the head sparsely covered with fuscous hairs; the labrum bears some long golden hairs; a straight keel runs from the base of the antennæ to the apex of the clypeus, the part on the latter being the thinner; the mandibles furrowed on the lower side; the teeth piceous; the apical two large. Antennæ black, shining, smooth. Thorax coarsely punctured; the scutellum more strongly than the mesonotum; the scutellum at the sides projecting into triangular teeth at the apex; the rest of the apex rounded, sparsely pilose; the edge of the pronotum and the base of the scutellum thickly covered with white hairs. The median segment sharply oblique, its sides thickly covered with white woolly hairs. Propleuræ finely and closely punctured; the mesopleuræ with the punctures much larger, those at the top being more widely separated and larger; the top, base, and apex thickly covered with white hairs. Legs thickly covered with white hairs, especially the tarsi; the base and apex of the fore femora; the hinder four femora entirely, the apex of the hind coxæ, the apex of the middle trochanters and the hinder trochanters, the hinder tibiæ and the hinder tarsi broadly at the base, rufous; the hinder tibiæ and base of tarsi strongly punctured; the lower part of the metatarsus covered thickly with long golden hairs; the calcaria pale golden. Abdomen closely punctured; the segments fringed with silvery hairs; interrupted on the basal segment; the abdominal segments strongly punctured; the scopa pale fulvous. Wings clear hyaline, the nervures and stigma fuscous.

MEGACHILE PARVULA, sp. nov.

Nigra, dense albo-pilosa; alis hyalinis, nervis nigro-fuscis. ♀. Long. fere 6 mm.

Hab. Mussouri (*Rothney*).

Front and vertex strongly punctured, the former not so strongly as the latter; the clypeus more closely and not quite so strongly punctured as the front; the vertex covered with long fuscous hairs; the inner orbits and the clypeus more thickly with white pubescence, especially the orbits; the clypeus slightly projecting; its apex curved; the labrum thickly fringed with golden hairs; the mandibles closely punctured; the apical three-fourths deeply furrowed; the two stout teeth rufous. Antennæ shining, almost glabrous. Thorax strongly punctured; the pronotum and the sides of the mesonotum fringed thickly with white hairs; the post-scutellum covered with much longer hairs. The base of the median segment with short stout longitudinal keels all over; those at the sides more widely separated and a little longer; the apex smooth, shining, impunctate, glabrous at the top; the rest of it obscurely punctured and sparsely covered with white hairs. Pleuræ coarsely punctured, and covered with long white hair; a shining, impunctate, semi-oblique furrow on the lower three-fourths of the mesopleuræ at the apex; the base of the metapleuræ finely and closely punctured. Legs thickly covered with white hairs; those on the under side of the hinder tarsi fulvous; the calcaria fulvous; the claws rufous. Wings clear hyaline; the nervures and stigma dark fuscous. Abdomen closely punctured; the base smooth and shining; the segments fringed with white hairs; the ventral scopa fulvous.

MEGACHILE CHRYSOGASTER, sp. nov.

Nigra; capite thoraceque albis; abdomine subtus longe aureo-hirto; alis hyalinis. ♀. Long. 7 mm.

Hab. Mussouri (*Rothney*).

Head strongly and closely punctured; the vertex sparsely, the front more thickly covered with long fuscous hairs; the sides of the clypeus and between the antennæ thickly covered with long white hairs; the mandibles coarsely punctured, their lower side fringed with long golden hairs. Antennæ shining, the scape slightly pilose; the flagellum glabrous. Mesonotum rather strongly punctured; sparsely covered with long fuscous hairs; the scutellum with the punctures larger and more widely separated; its apex fringed with long white hairs (longer than those on the mesonotum). The base of the median segment is coarsely crenulated, the edge being stoutly keeled; the apex with a sharp abrupt slope, and covered with long white hairs. Pleuræ strongly punctured, thickly covered with long white hairs. Abdomen shining; the basal segment with an abrupt, very slightly concave, slope; the top at the base with a distinct raised margin, the part behind this being depressed and crenulated. The other segments covered with shallow clearly separated punctures; the basal segments bare; the apical thickly covered with silvery pubescence, especially at their apices; ventral hairs dense, aureo-fulvous; the apical segment strongly punctured, fringed at the apex with pale fulvous hairs. Legs stout, covered with long white hairs; the anterior femora and tibiæ strongly punctured, the tibiæ almost rugose; the tarsi densely covered with long golden hairs. The wings clear hyaline; the nervures black; the tegulæ black, edged with white in front; shining, impunctate.

CERATINA PROPINQUA, *sp. nov.* (Pl. 4, f. 20).

Cærulea; clypeo, geniculis tarsisque albis; alis hyalinis.
♀. Long. 4—5 mm.
Hab. Mussouri (*Rothney*).

Head blue, an elongated shining impunctate white mark, rounded at the top, a little dilated at the apex, on the clypeus; the front in centre broadly carinate, the apex however of it being sharply margined, and from the ocelli a curved shining keel runs into it from either ocellus; the front and vertex very finely punctured; the face on either side of the white mark with large irregular punctures, which are continued up in a single row along the inner orbits. Antennæ black, brownish beneath. Mesonotum closely and finely punctured, shining; the scutellum more closely and finely punctured; the median segment finely and closely punctured, shining, except in the middle at the base; the pleuræ closely punctured. Legs thickly covered with white hairs; the anterior femora at the apex above and more broadly beneath, and the tibiæ and tarsi white; the tibiæ for the greater part fuscous behind; the hinder tibiæ fuscous, white at the base. Wings clear hyaline, the nervures fuscous, the stigma darker; the first and second transverse cubital nervures roundly curved and meeting closely at the top, where they are separated by about the space bounded by the first recurrent and second transverse cubital nervures; the third cubital cellule at the top is twice the length of the space bounded by the second recurrent and third transverse cubital nervures. Abdomen closely shagreened, shining and impunctate at the apices of the segments above and beneath.

Differs from the other green and blue species by the very much less strongly punctured head and thorax.

CERATINA TAPROBANÆ, sp. nov.

Viridis; abdomine nigro, late flavo-balteato; pedibus flavis; alis hyalinis. Long. fere 4 mm.

Hab. Ceylon (*Yerbury*).

Head green, shining, sparsely covered with pale hairs, clypeus, labrum, mandibles, and palpi pale yellow; the clypeus with two black lines. Scape of the antennæ yellow; the flagellum yellow beneath, black above. Thorax green, the scutellum dark purple; the metanotum of a darker green than the mesonotum, coarsely alutaceous; the pleuræ green, very finely punctured; the edge of the pronotum and the tubercles lemon-yellow. Legs yellow; covered densely with white hairs; the coxæ, trochanters, the basal third of the anterior femora; the basal half of the middle, and the basal fourth of the hinder, black. Wings hyaline; the stigma and nervures pallid testaceous; the recurrent nervures are almost interstitial. Abdomen pallid yellow; the second and third segments broadly black in the middle; the others broadly black at the base; the ventral segments broadly banded with black.

CERATINA BEATA, *sp. nov.*

Nigra, flavo-maculata; alis hyalinis. Long. fere 5 mm.
Hab. Trincomali, Ceylon (*Yerbury*).

Head shining, impunctate, glabrous, lemon-yellow; a broad black line runs down to the mandibles, curving round the side of the clypeus, and continued broadly upwards till it joins a broad black mark on the vertex, enclosing two yellow marks joined at the apex above the antennæ, these two yellow marks being joined to the lower yellow one by a narrow line between the antennæ; the inner orbits are yellow; behind the eyes is a broad yellow line which reaches near to the base of the mandibles, nearing the eyes as it does so; on the top of the occiput is a yellow line dilated in the middle; the vertex behind the eyes has large shallow punctures; the mandibular teeth black. Antennæ black, rufo-testaceous beneath; the scape having a yellower line. Prothorax yellow; the mesonotum black, with two yellow longitudinal lines in the centre and two narrower ones at the sides, neither reaching the base nor apex; its base obscurely punctured. Scutellum and post-scutellum yellow; the sides and apex of the median segment shagreened. Mesopleuræ closely punctured, yellow, except a broad line, below the wings and round the apex; the metapleura black; under the wings is a raised shining spot. Legs entirely lemon-yellow; thickly covered with white hairs. Abdomen yellow, shining, minutely punctured, the segments broadly black at the base; the ventral surface yellow.

CERATINA MODERATA, *sp. nov.*

Nigra; labro, clypeo, basi tibiarum, lineisque abdominis, flavis; alis hyalinis. Long. 5 mm.

Hab. Mussouri (*Rothney*).

Black; the labrum, clypeus, a curved line on the face close to the eyes and between the antennæ, the dilated yellow mark on the clypeus, and a transverse mark roundly dilated in the middle above, yellow. Front and vertex sparsely covered with long fuscous hairs; the face below the antennæ with three irregular rows of round punctures; the front depressed sharply between the antennæ and with a row of punctures between the furrow, the orbits above bordered with a row of punctures; the vertex punctured, except at the sides of the ocelli, the punctures large, clearly separated. Antennæ black, shining, the scape sparsely clothed with long white hair; the flagellum bare. Thorax shining, rather closely covered with white pubescence, which is darker in front than it is at the apex; the apex of the pronotum broadly yellow; the mesonotum shining, bearing widely separated shallow punctures; the scutellum minutely punctured. The base of the median segment has an elongated finely rugose area, bordered with a shining, impunctate space, the rest of the segment obscurely shagreened, and with an indistinct shallow narrow furrow down the middle; the propleuræ shining, the mesopleuræ with large, widely separated punctures; the metapleuræ alutaceous. Legs black, thickly covered with long white hairs; the lower apical half of the fore femora, the front four tibiæ before, the hind tibiæ also, but with a black line in the centre, bright yellow; the tarsi covered thickly with long white hairs; the front pair inclining to testaceous. Wings clear hyaline, the nervures testaceous; the first transverse cubital nervure is oblique; the second gradually curved towards it at the top; the second cubital cellule at the top is not half the length of the third; the first recurrent nervure is received slightly less than the length of top of the second cubital from the second transverse cubital nervure. Abdomen shining at the base, the rest more opaque, shagreened, and clothed closely with white hair; the ventral segments closely punctured.

CERATINA ORNATIFERA, *sp. nov.*

Long. 8 mm.

Hab. Mussouri (*Rothney*).

Very similar to *C. hieroglyphica*, but with the legs entirely black, except a yellow line on the tibiæ.

Head shining, sparsely covered with long fuscous hairs; the vertex with scattered punctures; a depression, narrowed towards the apex, in front of the anterior ocellus; the sides of which are flat, shining, and bear a row of punctures, there being a row of three inside of it at the apex; there is a sharp keel between the antennæ, from which the sides slope sharply and are punctured. The clypeus has widely separated punctures, except in the centre; the labrum is rugosely punctured. Antennæ entirely black, smooth, shining, the scape with a few long fuscous hairs. On the clypeus, extending from the base to the apex, is a broad yellow mark, which at its apex extends on either side to the base of the mandibles, where it is bounded by a curved furrow; on each side of the central yellow mark is a yellow mark with its apical half broadened on the inner side; and over the central yellow mark is a broad curved one which extends beyond its sides, and there are two oblique broad lines, narrowed at base and apex behind the eyes; above the antennæ are two yellow marks wider than long. Thorax black, shining; the pronotum with a broad line on either side in the centre; the tubercles, the apex of the tegulæ and a broad slightly curved mark on the scutellum, yellow; the scutellum slightly punctured at the base and apex; the median segment coarsely shagreened; the base flat, more shining in front laterally. The propleuræ at the top shining, smooth; the rest closely and strongly longitudinally-shagreened; the mesopleura and sternum with large distinctly separated punctures; the metapleuræ at the base closely punctured, the rest coarsely shagreened.

Legs thickly covered with long pale hairs; a mark on the under side of the fore femora; the tibiæ above yellow, the yellow becoming successively longer, but not reaching to the apex; the hairs on the hind tibiæ and tarsi very long and thick. Wings clear hyaline; the nervures and stigma fuscous; the top of the second cubital cellule is as wide as the space bounded by the first recurrent and the second transverse cubital nervures. Abdomen shining, smooth; the basal segment above, with a deep wide somewhat triangular depression, but with the apex rounded; on the apex of the basal segment is a yellow line, intersected by two square black marks, the second, third, and fourth segments bordered with yellow, which is dilated at the sides. Ventral segments strongly and closely punctured, lined with yellow at the apices and covered with long white hair.

The ♂ is similar; but with the yellow markings larger; there is a curved yellow mark on the mesonotum near the scutellum and two long narrow ones down the sides; the yellow line behind the eyes much larger; that on the scutellum much larger and dilated widely in the middle at the apex; the yellow on the anterior four tibiæ and tarsi more extended; and the yellow bands on the abdomen are broader above and beneath.

Anthophora Rothneyi, sp. nov.

Nigra, longe pallide hirta; abdomine pallide argenteo-fasciato; scapo antennarum, mandibularum basi oreque flavis; clypeo nigro-bimaculato; alis hyalinis, nervis nigris. ♀. Long. 10—11 mm.

Hab. Mussouri (*Rothney*).

Head densely covered with hoary hairs, longest and thickest on the top; the labrum and clypeus yellow, obscurely punctured; the two marks on the clypeus are dilated at the apex and reach from the base to shortly beyond the middle. Antennæ black, the scape yellow beneath, sparsely covered with long white hairs. Mandibles yellow, the apex black. Thorax covered, except on the scutellum, with long hoary hair; the scutellum closely and rather finely punctured, except in the middle. Legs black; the femora sparsely in front bearing white hairs; all the tibiæ in front and the four front tarsi in front thickly covered with silvery white hairs; the hinder tarsi thickly covered with black stiff hairs. Wings clear hyaline; the nervures black; the second cubital cellule at the top is as wide as the space bounded by the first transverse cubital and the first recurrent nervure; the second recurrent nervure is almost interstitial. Abdomen black; the ventral segments more or less piceous; the dorsal segments fringed with depressed silvery hair.

The ♂ wants the black marks on the yellow clypeus; the flagellum is brownish beneath; the apical ventral segment is deeply furrowed down the middle, and the sides are also deeply and more widely furrowed; the bounding keels being acute, stout; the space between them in the centre slightly hollowed.

Agrees with *A. cincta* in form and general coloration; but is smaller; the hairs on the head and thorax hoary, not fulvous; and the fasciæ on the abdomen are pale silvery, not blue.

SOCIALES.

TRIGONA BENGALENSIS, *sp. nov.*

Nigra, nitida; antennis flavo-testaceis; abdominis basi late brunneo; alis hyalinis. ♀. Long. 3 mm.

Hab. Barrackpore (*Rothney*), in old tree stumps.

Head black, shining, the face thickly covered with white pubescence, the mandibles and palpi rufo-testaceous. Antennæ entirely rufo-testaceous. Thorax shining, impunctate; the pleuræ thickly covered with long white hairs, but more sparsely in the middle; the mesonotum sparsely covered with longish fuscous hairs; the semicircular depression at the base of the scutellum deep; the scutellum fringed with long fuscous hairs at the top and at the apex behind, the latter oblique, projecting at the top; the median segment very smooth, shining, glabrous, rounded. Legs covered with white hairs, shining, the apical four joints of the tarsi testaceous; the hinder femora and tibiæ have a piceous tinge. Wings clear hyaline, the stigma pallid-testaceous; the radial nervure complete; the cubital only extending to the middle of the second cubital cellule; the two transverse cubital nervures very faint, almost obsolete.

Explanation of Plates.

PLATE 3.

Fig. 1. *Ichneumon clotho.*
,, 2. *Ichneumon Rothneyi.*
,, 3. *Rothneyia Wroughtoni.*
,, 4. *Pimpla nepe.*
,, 5. *Bracon ceylonicus.*
,, 6. *Bracon agraensis.*
,, 7. *Spinaria nigriceps.*
,, 8. *Epyris amatorius.*
,, 9. *Chalcis bengalensis.*
,, 10. *Temnata maculipennis.*

PLATE 4.

Fig. 11. *Methoca ♂ rugosa.*
,, 12. *Methoca ♀ bicolor.*
,, 13. *Mutilla œdipus.*
,, 14. *Mutilla Rothneyi.*
,, 15. *Pison (Parapison) Rothneyi.*
,, 16. *Cemonus fuscipennis.*
,, 17. *Rhynchium basimacula.*
,, 18. *Halictus tardus.*
,, 19. *Andrina gracillima.*
,, 20. *Ceratina propinqua.*

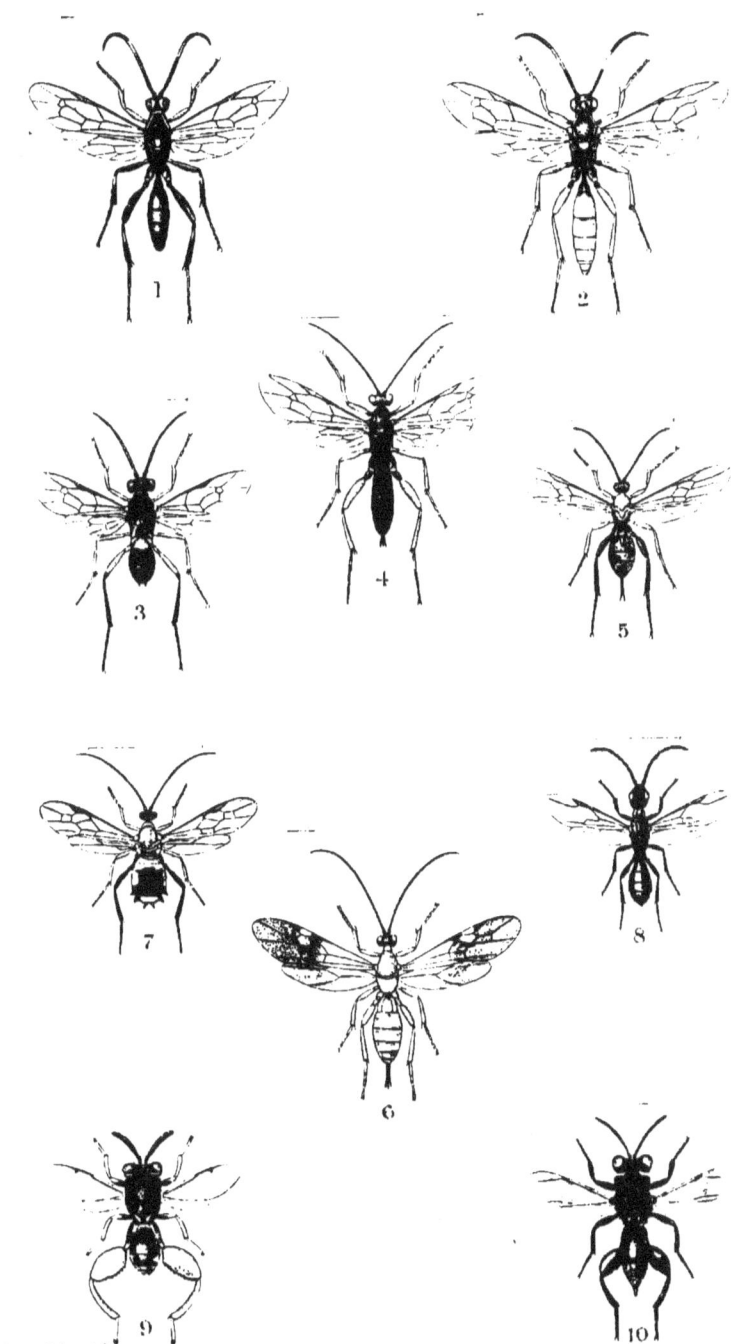

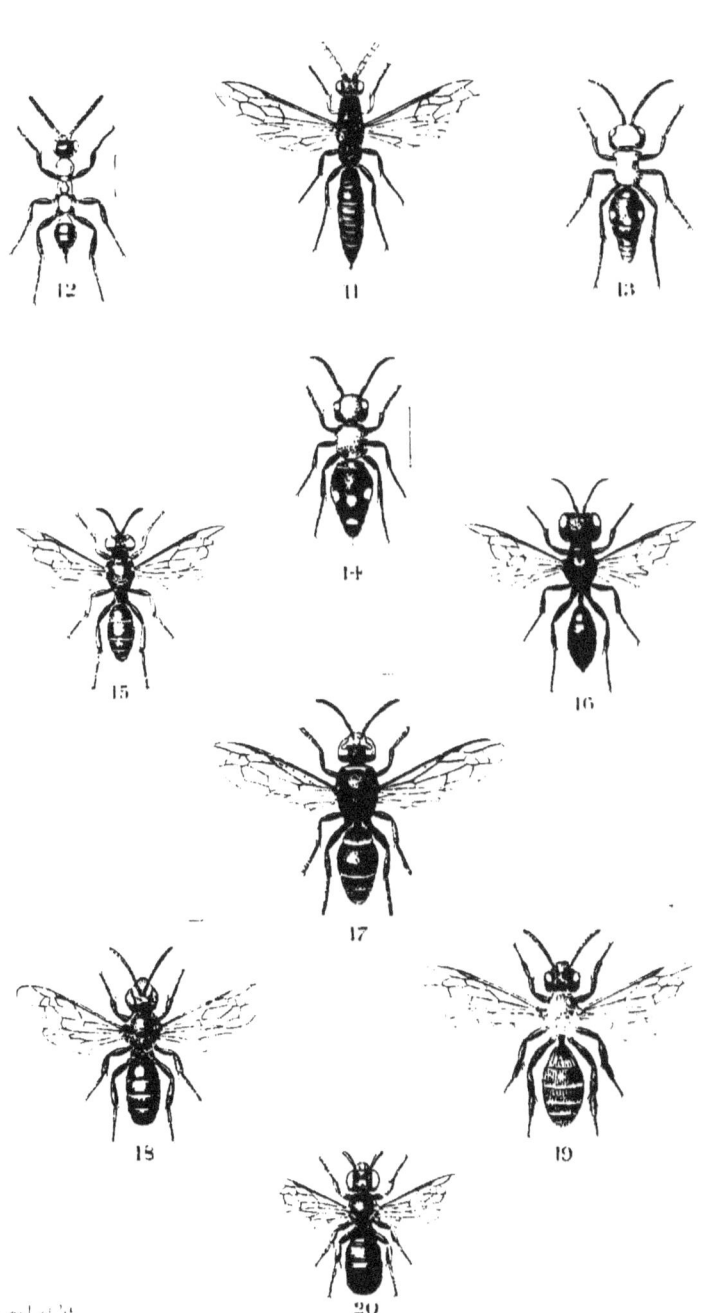

[From Volume 41, Part IV., of "MEMOIRS AND PROCEEDINGS OF THE MANCHESTER LITERARY AND PHILOSOPHICAL SOCIETY," Session 1896-7.]

Hymenoptera Orientalia, or Contributions to a knowledge of the Hymenoptera of the Oriental Zoological Region. Part VI.

BY

PETER CAMERON.

209033

MANCHESTER:
36, GEORGE STREET.

1897.

XIII. Hymenoptera Orientalia, or Contributions to a knowledge of the Hymenoptera of the Oriental Zoological Region. Part VI.

By Peter Cameron.

[*Communicated by J. Cosmo Melvill, M.A., F.L.S.*]

Received March 16th. Read March 23rd, 1897.

In continuation of my last paper on Indian Hymenoptera, I give here descriptions of new species belonging to various groups.

CHRYSIDIDÆ.

This interesting family is being revised, as regards the Indian species, by M. Robert du Buysson, in the *Journal of the Bombay Natural History Society*, from material chiefly collected by Mr. R. C. Wroughton, in Bengal.

Chrysis.

A. Apex of abdomen undulated, without distinct teeth.

Chrysis perfecta, *sp. nov.*

Long. 8—9 mm.

Hab. Barrackpore.

Green, the flagellum of the antennæ black; the vertex, the middle of the pronotum and mesonotum, bluish; the apex of the first abdominal segment narrowly, of the second segment to near the middle, bluish, with a purplish tinge; the tarsi blackish, the basal joint testaceous; the wings fuscous, lighter coloured at the apex.

September 22nd, 1897.

The scape of the antennæ, the second joint, and the base of the third, green; the flagellum black, covered sparsely with a microscopic white down; the third joint slightly, but perceptibly, longer than the fourth. The vertex coarsely punctured; a curved keel runs from the side of the lower ocellus, the space between being irregularly longitudinally striolated, except in the centre at the apex, where it is smooth; the space below the antennæ thickly covered with long white hair; the clypeus has some scattered, moderately large punctures; the mandibles black, piceous towards the middle. Pronotum coarsely punctured; depressed in the middle; the mesonotum with the punctures as large, but, if anything, more widely separated; the median segment with the punctures still larger, deeper, and coarser, and with an oblique wide depression at the sides, this depression having at the base a few stout oblique keels. Propleuræ strongly and somewhat irregularly punctured, and with a large oval depression on its lower side; the mesopleuræ in front with small punctures; the rest with the punctures large, coarse, and deep; at the bottom are five stout perpendicular keels, which form one oval and three straight foveæ; the sternum finely punctured; the mesosternum with a straight keel, which does not quite reach to the apex. Legs green; the femora and tibiæ punctured, sparsely covered with white hair; the tarsi with the hair shorter and thicker, black; the base broadly testaceous. Wings fuscous, with a violaceous tinge; the apex almost hyaline. Abdomen large; the puncturing moderately close, deep, and of about equal strength on all the segments; on all closer and stronger laterally. The apical segment waved; without teeth, but with a rounded projection on either side of the middle; there are 10 foveæ, the central large, broader than long, the others smaller and rounder; the ventral surface green, finely punctured, the apical segment with a longitudinal furrow down the centre; the apex blackish.

B. Apex of third abdominal segment rounded, with a tooth on either side at the end.

CHRYSIS FURIOSA, *sp. nov.* (Pl. 16, f. 9).

Viridis, late cæruleo-maculata, flagello antennarum nigro; alis fere hyalinis. ♀. Long. 6 mm.

Hab. Barrackpore.

Head nearly as wide as the thorax, green, a large purplish-blue patch on the vertex. On the front distinctly below the ocelli is a stout keel depressed inwardly, broadly in the centre, continued as a straight somewhat oblique keel down the sides of the eyes, from which it is distinctly separated, and becoming united to a semicircular keel over the frontal depression, which is finely and closely punctured at the sides, and finely transversely striated in the middle. From the centre of the top frontal keel a short thick keel runs down to the centre of the area. The vertex bears longish fuscous hair; the sides of the frontal depression with short white hair. Antennæ black; the scape, second and third joints, green; the flagellum covered with a microscopic down; the scape with white hair; the third joint is fully one quarter longer than the fourth. Prothorax half the length of the head; the sides, seen from above, oblique at the base; the rest curved, the base and apex of the curve being distinct, almost forming teeth; the punctuation coarse and even; in the central region is a large bluish band. Mesonotum coarsely punctured, the punctures large, round, and deep; the punctuation on the scutellum and the middle of the median segment similar to that on the mesonotum. The front of the propleuræ coarsely punctured; the rest excavated, shagreened; the mesopleuræ coarsely punctured, almost running into reticulations; the hinder part with the punctuation closer and finer; over the sternum is a wide, deep, longitudinal furrow. The prosternum has a semicircular furrow at the apex; and down its centre, before and behind the semicircular furrow, is a straight keel.

C. Apex of abdomen with a tooth on either side and in the middle.

This section resembles closely that of the preceding, from which it differs in having a tooth in the centre of the apex of the abdomen, and also in the front having no area enclosed by stout keels.

CHRYSIS MENDICALIS, *sp. nov.*

Long. 6 mm.

Hab. Barrackpore.

Green, the centre of the thorax and the greater part of the second and third abdominal segments, blue; the coxæ, femora, and tibiæ green; the tarsi fuscous; the wings almost hyaline.

Antennæ stout, black; the first and second joints green; the third joint hardly one-quarter longer than the fourth. Vertex coarsely punctured, almost reticulated on the lower side; the ocellar region, and the part behind, bluish-purple; the excavated front closely punctured at the side; the centre finely transversely striated; the keel over the depression stout, and reaching to near the bottom of the eyes. Pronotum slightly more than half the length of the eyes; the sides oblique, rounded in front, uniformly coarsely punctured, the central area more coarsely, almost running into reticulations towards the apex, and for the greater part purplish. The propleuræ with a large oval depression at the top, which is almost smooth; the lower part shagreened. The mesopleuræ coarsely punctured, almost reticulated; a small basal and a larger apical, smooth, deep depression on the side of the mesosternum; the metapleuræ almost smooth above, shagreened beneath. Wings almost hyaline; the nervures fuscous; the stigma black; the tegulæ green, punctured. Abdomen with a triangular depression in the centre at the base; the basal segment green, tinged with blue; the second and third segments blue, their sides green; the apices with a brassy tinge; there are on each side of the centre two distinct foveæ of nearly equal size, and a much larger one looking like two united, at least on one side of the example examined; for on the other side there are four foveæ, all clearly separated from each other. Ventral surface smooth, shining, green, with brassy tints.

D. Apex of abdomen with four teeth.

CHRYSIS DISPARILIS, *sp. nov.*

Long. nearly 7 mm.

Hab. Barrackpore.

Green; the vertex, the mesonotum, and the greater part of the abdominal dorsal segments, bluish-purple; the tarsi broadly testaceous at the base.

Head as wide as the mesonotum and somewhat longer than the pronotum, green; the ocellar region and the greater part of the occiput purple, coarsely punctured: the cheeks thickly covered with long white hair. Over the frontal depression and below the ocelli is a curved keel. Antennæ stout; the third joint on the lower side equal in length to the fourth. The thorax strongly punctured; the pronotum entirely green; the mesonotum purple; the scutellum and metanotum green; the scutellum with a faint purplish tinge. Wings hyaline. The first abdominal segment green, with a large purplish mark across the middle; the second segment purple, green at the apex; the last purple, green before the teeth, which are themselves purple.

Hedychrum lugubre, sp. nov. (Pl. 16 f. 10).

Cæruleum, viride maculatum; flagello antennarum fusco; tarsis fusco-testaceis; alis fuscis. Long. 4—5 mm.

Hab. Barrackpore.

Scape of the antennæ green, rather strongly punctured; the flagellum fuscous-black, covered with a pale fuscous down; the third joint about one-quarter longer than the fourth. Head strongly punctured, the frontal depression finely and closely transversely striated. The base of the mandibles green, the rest of them blackish, with a piceous stripe before the middle. Prothorax longer than the head, strongly punctured above; two large blue splashes, nearly united, behind; its sides oblique, except at the apex, which is straight. Mesonotum in the middle purplish; the base, especially in the centre, with the punctures smaller and more widely separated than they are at the sides, and more especially behind; the scutellum, with the punctures larger and deeper than they are on the mesonotum; the median segments with the punctures larger and deeper than on the scutellum. Propleuræ in front coarsely punctured; behind smooth; 'the mesopleuræ coarsely punctured, behind finely striated; the metapleuræ finely longitudinally striated, more finely on the lower than on the upper side; the mesopleuræ finely transversely striated. Legs green, the tibiæ with a bluer tinge than the femora; the apex of the tibiæ testaceous; the tarsi dark testaceous, paler towards the apex. Wings uniformly dark fuscous; the nervures darker. Abdomen bluish, greener in tint in the middle; the punctures on the second segment more widely separated than on the basal; that on the third much coarser than on either. The ventral surface smooth, sparsely covered with white hair.

The head and thorax may have brassy tints; the relative proportions of the blue and green in the abdomen varies.

ICHNEUMONIDÆ.

ICHNEUMON ARDATES, sp. nov.

Niger, scutello flavo; abdomine rufo, flavo- et nigro-balteato; pedibus rufis; alis fere hyalinis, stigmate flavo. ♀. Long. 13 mm.

Hab. South India.

Head coarsely punctured; the inner orbits in the middle broadly, and a mark immediately below the antennæ, rufous; the palpi pale yellow. Antennæ short, thick, the 13—14 basal joints pale yellowish-testaceous, the apices deep black. Thorax black, except the scutellum which is pale yellow; closely punctured; the propleuræ closely obliquely striated except at the top. The lower part of the mesopleuræ coarsely punctured; the upper shining, irregularly shagreened; the metapleuræ closely coarsely irregularly striated. Median segment closely rugose; only the supramedian area clearly defined; it is a little wider than long, slightly narrowed from the middle; the apex in the middle a little dilated inwardly. Legs, including the coxæ, rufous. The narrow part of the petiole rufous; the broad apex with a yellowish band, the sides black; the raised central part of the apex closely longitudinally striated. The sides of the gastrocœli stoutly striated. The second segment is entirely rufous; the basal half of the third segment and the whole of the fourth black; the apical half of the third yellow, with a reddish tinge; the apical segments pale-yellow.

Ichneumon Ælvanus, sp. nov.

Capite, abdomine pedibusque rufis; thorace nigro, rufo-maculato; antennis rufis, apice late nigro; pedibus rufis, coxis posticis nigris; alis fusco-violaceis. ♂. Long. 23 mm.

Hab. India, South-east Provinces.

Head entirely rufous, strongly punctured all over; the orbits distinctly margined on the inner side: the vertex broadly depressed; the tips of the mandibles black. Antennæ stout, the flagellum bare, from the 15th joint deep black. Thorax strongly punctured, black; the pronotum in front in the middle, its sides, a mark in front of the scutellum, the scutellum and post-scutellum, the tubercles and the lower part of the mesopleuræ in front, rufous; median segment strongly punctured, except in the middle at the base; the supramedian area rounded and narrowed at the base. Legs closely covered with short hair; the tibiæ and tarsi are paler than the femora. Wings uniformly fuscous violaceous; the areolet longish, narrow; at the top slightly narrower than the space bounded by the recurrent and second transverse cubital nervures. The basal half of the petiole black; coarsely punctured, especially toward the apex; the sides there depressed, and there is near the base of the dilated part a shining, impunctate, somewhat triangular space; the gastrocœli longitudinally striated from the base to near the apex; the striæ stout, all distinctly separated and of nearly equal thickness. The apical segments are thickly covered with short fulvous hair.

Ichneumon Godwin-Austeni, sp. nov.

Cæruleus, orbitis oculorum, geniculis tibiisque anticis flavis; alis fusco-violaceis. ♀. Long. 15 mm.

Hab. Khasi *(Godwin-Austen).*

Antennæ black, the 11th and 12th joints white beneath and at the sides, from these joints becoming thickened and slightly compressed laterally; almost bare. Head shining, the outer orbits at and a little below the middle and the inner from the top to shortly below the middle, yellowish. Face flat, slightly dilated in the middle below the antennæ; punctured, the punctures at the sides more widely separated than in the middle, this being also the case on the front and vertex. Pro- and meso-thorax closely and uniformly punctured; the scutellum more shining and sparsely punctured; the median segment more strongly and closer punctured than the mesonotum, which has the areæ all clearly defined. Legs blue, the coxæ closely and thickly covered with white hair; the femora sparsely haired. Areolet narrowed at the top, the transverse cubital nervures being almost united; the recurrent nervure is received in the apical third of the areolet. Abdomen closely and strongly punctured; the apex of the petiole raised; the raised part clearly margined and separated from the sides; this raised part is longitudinally striated. Gastrocœli wide, deep; their sides obliquely striated; the part between the gastrocœli in the middle longitudinally at the sides, especially at the base, more strongly obliquely, striolated.

This species has the antennæ thickened towards the apex somewhat as it is in *Joppa*, but I cannot look upon it as congeneric with the American species of *Joppa*. Apart from the dilated antennæ (and in this point we find considerable variation in *Ichneumon*) I can find nothing whereby to distinguish it from *Ichneumon* as generally defined.

CRYPTUS PERPULCHER, *sp. nov.* (Pl. 16, f. 7).

Niger, thorace abdomineque flavo-maculatis; pedibus fulvis, apice femorum posticorum basique tibiarum posticarum, nigris; alis hyalinis. ♂. Long. 13 mm.

Hab. Borneo.

Antennæ black, bare, a broad white band near the middle. Head black; a mark close to the eyes opposite the ocelli, a mark touching the eyes immediately under the antennæ, and having a somewhat roundish projection issuing obliquely from above its middle, the mandibles (except the teeth) and the palpi, whitish-yellow; the face above the clypeus irregularly striated; front depressed, especially over the antennæ, where there is a straight keel in the centre; below the ocelli are a few oblique stout striæ. Pro- and meso-notum smooth, shining, impunctate; the middle lobe of the mesonotum well developed; the pro- and meso-pleuræ stoutly longitudinally striolated, except the former above at the base, where there is a smooth triangular spot; and the apex of the mesopleuræ where the striations only extend to the lower side. The base of the median segment before the keel is smooth, the rest of it closely transversely striated; the metapleuræ irregularly rugosely punctured. The following yellow marks are on the thorax: The tegulæ, scutellum, post-scutellum, an elongated triangular mark before the spiracles; a smaller one in front of and above the hinder coxæ; and a ⊥-shaped mark on the median segment extending from near the transverse keel to the apex of the segment, the cross piece being thicker than the longitudinal. The coxæ are black, except the anterior at the extreme apex; the middle pair have a small mark at the base, and the hinder part a large yellow mark extending from the base to near the apex, its inner end being more prolonged than its outer; the fore coxæ and trochanters are whiter in tint than the others; the hinder trochanters are black, this being also the case with the apex of the femora, and to a less extent the base of the tibiæ; the four front tarsi are dark fuscous; the apex of the hind tibiæ, the base of the metatarsus broadly, and the extreme apex of the tarsi, black; the fore tarsi are infuscated; the middle almost black. Wings hyaline, the stigma and nervures black; the areolet longer than broad; the second recurrent nervure received near the apical third of the cellule. Abdomen black, shining. All the segments pale yellow at the apex; the apical one almost entirely yellow; the ventral segments black, dull yellow at the apices.

CRYPTUS BROOKEANUS, *sp. nov.*

Niger, annulo late flagello antennarum tarsisque posticis albis; pedibus fulvis; trochanteribus posticis, apice femorum posticorum tarsisque posticis nigris; alis hyalinis. ♀. Long. 11 mm.

Hab. Borneo.

Antennæ as long as the body, from the apex of the fifth to the base of the fourteenth joint white above and at the sides; bare. Head black, bearing a short, sparse, black pubescence; the front, except at the sides, irregularly longitudinally striolated: the striæ rather stout; the face shagreened; the palpi white. Thorax entirely black; the pro- and meso-notum almost shining, impunctate; the base of the median segment behind the transverse keel (which is broadly curved backwards in the centre) irregularly longitudinally striated, and with two stout straight keels down the centre; behind this keel it is irregularly reticulated; in the middle at the sides are two stout spines. The upper part of the propleuræ is obliquely striolated, the striæ becoming stronger towards the apex, at the base in the middle being almost obsolete; the portion over the coxæ impunctate, smooth; the mesopleuræ closely irregularly longitudinally striolated; immediately under the wings are a few stout, clearly-separated oblique striations, which are mostly turned up at the base. The mesosternum smooth, impunctate, except a crenulated furrow down its centre, and separated from the pleuræ by a curved crenulated furrow. Legs fulvous; the four anterior tibiæ and tarsi infuscated; the fore tibiæ white in front; the hinder trochanters, the apical third of the hinder femora and the hinder tibiæ, black; the hinder spurs black; the tarsi white, the extreme apex black. The second transverse cubital nervure is faint; the recurrent nervure is received in the apical third of the cellule. The petiole is smooth and shining; the sides at the apex depressed; the rest of it alutaceous; the apex of the second segment pale testaceous.

ICHNEUMON MITRA, *sp. nov.* (Pl. 16, f. 6).
Long. 12 mm.
Hab. Borneo.

Antennæ stout, almost bare, the 10—15 joints white except beneath. Head black; the orbits from shortly above the antennæ to shortly behind the hinder ocelli, the sides and apex of the clypeus (the latter narrowly), the labrum and palpi, yellowish-white; the mandibles piceous before the middle. The face rather strongly punctured; the punctures on the clypeus more widely separated towards its apex; the space above the antennæ shining and impunctate: the vertex coarsely punctured. The face and vertex covered with short white hair; the inner orbits distinctly margined. Thorax black; the edge of the pronotum, tubercles, tegulæ in front, and scutellum, yellowish-white. Thorax closely punctured; the propleuræ closely punctured above, beneath shining, and with strong, somewhat oblique keels behind; the mesopleuræ with the punctures larger and more clearly separated above, the lower part with them smaller and much more closely set together, and at the apex running into striæ. The mesonotum closely punctured; the scutellum has the punctures shallower and more clearly separated; the yellow mark does not occupy quite its entire surface, and is somewhat mitre-shaped; the post scutellum shining, impunctate. The median segment strongly punctured, thickly covered (especially behind) with white hair; the supramedian area longer than broad; the sides almost straight; the base transverse; the apex curved roundly inwardly. The metapleuræ are more coarsely punctured than the mesopleuræ, and more thickly haired. Legs black; the front tibiæ and tarsi dirty testaceous (perhaps discoloured); the outer half of the fore coxæ, the outer side of the middle and a larger mark on the hinder side of the hinder coxæ, yellowish-white; the spurs also yellowish-white. Wings hyaline; the stigma and nervures black; the latter paler towards the apex; the areolet at the top in length a little less than the space bounded by the recurrent and the second transverse cubital nervures. Abdomen black; the base of the first, second, and third segments with yellow bands dilated at the sides, and which become gradually narrowed, a large mark on the sixth, rounded at the base, narrowed gradually at the sides and the greater part of the seventh, yellowish-white. The ventral segments black; the basal yellowish in the middle.

This species is abundantly distinct from the two species of *Ichneumon* described by Smith from Borneo, the antennæ of *I. penetrans* and the head of *I. comissator* being for the greater part yellow.

BRACONIDÆ.

BRACON BORNEENSIS, *sp. nov.*

Capite, thorace pedibusque flavis, abdomine ferrugineo; alis fuscis, basi late flavo; flagello antennarum nigro. ♂. Long. 10 mm.

Hab. Borneo.

Scape of antennæ pale yellow, sparsely covered with longish hairs; the flagellum entirely black. Head shining, sparsely covered with fuscous hair, which is longer and paler below the antennæ than on the vertex. At the sides and behind the ocelli are bordered by a distinct furrow; in front of them is a depression from which a straight narrow furrow runs to the base of the antennæ. Thorax smooth, shining, impunctate, the upper part fulvous, the sides and sternum paler. Legs fulvous, sparsely haired. Wings from the transverse basal nervure dark fuscous, with a slight fulvous tinge; the stigma and the nervure in the fuscous part of the wings, blackish; in the yellowish, yellow. The central part of the petiole above has a few widely separated keels; the lateral furrows wide and deep; the central part has the sides at the apex rounded and with a wide short furrow on the inner side. Down the centre of the second segment is a straight keel depressed in the middle, and which does not quite reach the apex of the segment; on either side at the base is a wide oblique depression reaching near to the apex; at its base are some sharply oblique keels; the rest of it has a few semi-oblique keels; the suturiform articulation has throughout straight stout keels; the other segments coarsely rugosely punctured, except the last, which is smooth, shining, impunctate, and of a pale yellow colour.

Is not unlike *B. Rothneyi* but is larger, the base of the wings more broadly yellow; the lateral depression on the second abdominal segment is much larger and deeper; the central keel much more complete and clearly defined and without a smooth triangular base; the base of the antennæ yellow, not black, &c.

BRACON DISSIMULANDUS, sp. nov.

Niger; capite, pro- mesothorace metapleurisque ferrugineis; alis fuscis, fere violaceis. ♀. Long. 14; terebra 5 mm.

Hab. Borneo.

Head ferruginous, the teeth of the mandibles black; rather closely covered with black hair, particularly on the face; the front and vertex smooth, the face smooth in the centre, the sides with large, shallow, distinctly separated punctures; the sides of the clypeus have a yellowish hue; the palpi are covered with long, black hair; the mandibular teeth black. Antennæ entirely black; the scape with longish black hair. Pro- and meso-thorax smooth and impunctate: their pleuræ and sternum sparsely covered with fuscous hairs. The metathorax thickly covered with longish black hair; the upper part almost entirely black. The two anterior legs entirely ferruginous; the four hinder black; the intermediate with the base of the coxæ broadly, and the extreme base and apex of their femora ferruginous; the hinder legs are thickly haired. Wings large, uniformly smoky-violaceous; there is an elongated clear hyaline spot below the first transverse cubital nervure. The petiole is deeply depressed at the base; the raised centre bordered along the sides by a wide, moderately deep, shallow furrow: the raised central part bearing stout longitudinal keels: the central being stouter and straighter; at the apex of the segment there are shorter keels between the longer ones, or those become bifid. The third segment is nearly similarly striolated, but with the striæ closer together: and there is at the apex an interrupted transverse furrow: the remaining segments shining, smooth; the ventral surface pale-yellowish, the sheaths of the ovipositor thickly covered with long hair.

Of the Oriental species it comes nearest to *B. foveatus* Sm., but that has the ovipositor twice the length of the body.

BRACON CHARAXUS, *sp. nov.*

Niger: capite, thorace pedibusque anticis ferrugineis; alis fuscis, fere violaceis. ♀. Long. 11; terebra 12 mm.

Hab. Borneo.

Antennæ black; the flagellum almost bare; the scape thickly covered with blackish hairs, and piceous in the middle beneath. Head shining, sparsely covered with long fuscous hairs; below the antennæ bearing all over except in the middle, where there is a smooth space, moderately large punctures; the front and vertex very smooth and shining, except for a few small punctures along the inner orbits; the mandibles ferruginous, their teeth black; the palpi fuscous. Thorax entirely ferruginous, smooth, shining, impunctate; the middle lobe of the mesonotum raised; the median segment sparsely covered with long black hairs; on its side is a deep wide furrow, which does not reach the base. Wings uniformly deep smoky, but with a violaceous tinge. The fore legs entirely ferruginous, as are also the middle coxæ, except that they are darker; the four anterior legs sparsely covered with short hair: the hinder tibiæ and tarsi have the hair much longer and thicker. The petiole above is smooth and shining, except the apex in the middle, where it is a little rough; the middle part bounded by the keels is almost transverse, and has behind it a small space bearing some minute punctures; the outer divisions at the apex are obliquely truncated. The 2—4 segments are closely and strongly longitudinally striolated, the striations on the second being irregular; in the centre of the second segment are two keels, which unite in the middle and are continued to the apex as one: from the base near the edge runs another keel, which runs obliquely to the central keel, when it becomes straight; the sides are distinctly margined above; on the side of the second segment is a large smooth, shining space; and there is a similar one, but smaller, on the third in front of the depression; the other segments are smooth and shining, the last is depressed at the base and is fringed at the apex with longish hairs. The ventral segments, except at the apex, are, in the middle, yellowish-testaceous; the last ventral segment projects beyond the apex of the dorsal.

Allied apparently to *B. foveatus* Sm. from Singapore; but that, among other differences, has the ovipositor twice the length of the body.

POMPILIDÆ.

SALIUS LEPTOCERUS, sp. nov.

Niger, abdomine pedibusque rufis; capite, thorace coxisque dense fulvo-hirtis; alis fusco-violaceis. Long. 17; exp. al. 24 mm.

Hab. Sikim.

Antennæ a little longer than the body, entirely black, except the scape on the under side, which is rufous. Head densely covered all over with a golden fulvous pile and less densely with long fulvous hair. Mandibles densely covered with short depressed fulvous pubescence; the palpi blackish. Thorax densely covered all over with golden fulvous pubescence and more sparsely with longish pale fulvous hair; there is a wide, deep furrow down the centre of the post-scutellum, and there is a narrower, less distinct one down the base of the median segment, which is obscurely transversely striated. Wings fuscous-violaceous, shining, the nervures blackish; the first recurrent nervure is received a short distance in front of the second transverse cubital; the nervures dark fuscous, the stigma darker at the base. Legs red; the coxæ black, densely covered with golden pubescence and, more sparsely, with longish fulvous hair; the posterior are rufous on the under side; the trochanters are black at the base. Abdomen dark fulvous; the second, third, and fourth segments black at the base.

Comes near to *S. zelotypus* Bingham from Tenasserim.

DOLICHUSUS CLAVIPES, sp. nov. (Pl. 16, f. 4).
Niger ; alis hyalinis. ♀. Long. 9 mm.
Hab. Barrackpore (Rothney).

Antennæ filiform. Immediately above, and slightly protruding over them is a large projection which, seen from the side, is triangular; above depressed, the sides and apex distinctly raised; the base not margined; the front and vertex shining, impunctate. Antennæ separated from the base of the clypeus, which is keeled down the centre. Eyes reaching to the base of the mandibles. Radial and cubital cellules not differing from *Pseudagenia*. Prothorax somewhat longer than in typical *Pseudagenia*. Mesonotum with two nearly complete, deep parapsidal furrows; the median segment with distinct areæ; on the sides on the top of the apical part is a small blunt tooth, and in the middle is a much larger and more distinct one. At the base of the third ventral segment is a transverse furrow; the sheath of the ovipositor largely projecting. Claws with one tooth.

Antennæ filiform, the scape sparsely haired; the flagellum closely covered with a short pubescence. Head shining, impunctate; sparsely haired; the outer orbits on the lower side thickly covered with longish white hair; the clypeus, especially at the sides, and the base of the mandibles with longer white hairs. Thorax shining; pro- and meso-notum thickly covered with fuscous hair; the hair on the median segment longer and thicker; the apex of the pronotum depressed and clearly separated from the mesonotum. The parapsidal furrows do not quite reach to the apex of the mesonotum. Apex of scutellum semicircular; post-scutellum stoutly longitudinally striolated. In the centre of the median segment are two keels which converge a little at the apex of the flat part, and these are united by a transverse keel; the centre at the base shagreened and with four irregular longitudinal keels; the oblique apex shagreened. Propleuræ shining; the mesopleuræ shining above, shagreened below; the top projecting, oblique; a keel runs down the base from the tubercles; the metapleuræ closely longitudinally striolated. Legs shining, sparsely haired; the base of the hind spurs thickly covered with stiff pale hairs. Wings clear hyaline; the nervures blackish; the first transverse cubital is sharply elbowed from a little below the middle towards the apex of the cellule, the second straight, the third curved roundly toward the base of the cellule; the first recurrent nervure is received shortly beyond the middle of the cellule, the second near the basal third.

SPHEGIDÆ.

DIODONTUS STRIOLATUS, *sp. nov.* (Pl. 16, f. 3).

Niger, mandibulis, tegulis, geniculis, tibiis tarsisque flavis: alis hyalinis, nervis stigmateque fuscis. ♂. Long. fere 5 mm.

Hab. Lahore (*Rothney*).

Antennæ entirely black, almost bare; the apex of the scape fuscous. Head shining, the front and vertex with fine, distinctly-separated punctures; mandibles yellow, the extreme base black, the teeth piceous-black; the palpi yellow; the clypeus projecting, roundly and deeply incised in the middle. Thorax shining, faintly aciculated above; the propleuræ with stout, distinctly separated striæ; the apical half of the mesopleuræ closely longitudinally striated, the striations becoming closer together at the apex; the metapleuræ, except at the base beneath, more strongly and irregularly striolated. The apex of the four front femora, the tibiæ and tarsi, testaceous; the middle tibiæ infuscated behind; the hinder tibiæ blackish; the hinder tarsi infuscated. Wings short, not reaching much beyond the middle of the abdomen, slightly infuscated, the nervures testaceous, the stigma black; at the top the second cubital cellule is slightly wider than the space bounded by the first transverse cubital and the second recurrent nervures. Legs sparsely covered with white pubescence; the apex of the femora, tibiæ and tarsi, testaceous; the hinder tibiæ infuscated. Abdomen shining.

DIDINEIS ORIENTALIS, *sp. nov.* (Pl. 16, f. 2).

Niger, mandibulis, scapo antennarum subtus, tibiis, tarsis tegulisque albidis, alis hyalinis, nervis fuscis. ♂. Long. 5 mm.

Hab. Barrackpore (*Rothney*).

Antennæ fuscous, darker above, the scape bearing a few hairs, the flagellum thickly covered with short pile; the base of the apical joint before the base of the curve projecting. Head shining, the vertex with shallow closely-pressed punctures, and covered with longish blackish hair; the vertex with the hair shorter and closer; the cheeks and clypeus thickly covered with silvery hair, that on the clypeus being the longer. Mandibles with longish silvery hair; their base black, the rest piceous, with a yellow band between: the palpi yellow. Thorax black, shining, closely covered on the pro- and meso-thorax with black hair; almost impunctate; the depression on the propleuræ with a few stout, oblique keels. In the centre of the median segment is a large somewhat triangular area, but with the apex rounded, bounded by stout keels, and having in the centre of it two slightly diverging keels, which reach a little beyond the middle. From the apex of the triangle a straight keel runs down to the apex of the segment, and in the centre at the side is a somewhat semicircular area, which is joined to the central keel by two short transverse ones. Legs thickly covered with short white hair; the apices of the coxæ, of the trochanters and of the femora, the base of the hinder tibiæ, the four anterior tibiæ and all the tarsi, yellowish-testaceous; the femoral tooth stout, oblique, somewhat triangular; the apex of the hinder femora fuscous. The wings have a faint fuscous tinge; the stigma fuscous, the nervures dark testaceous; the second cubital cellule oblique, at the bottom longer than the third cellule; the recurrent nervures almost interstitial; the third transverse cubital nervure is curved at the top; the lower part straight, oblique. Abdomen smooth and shining, sparsely covered with longish white hair, which becomes longer and thicker towards the apex.

This and *Alyson* are interesting additions to the Oriental Zoological Region, the few described species being from Europe and North America.

ALYSON ANNULIPES, *sp. nov.* (Pl. 16, f. 1).

Niger, mandibulis basi tibiarum posticarum maculisque 2 abdominis flavis; alis hyalinis, fusco-fasciatis. ♀. Long. 6 mm.

Hab. Poona (*Wroughton*).

Black, shining, almost impunctate; the head, pro- and meso-thorax sparsely covered with long fuscous hair; the apex of the median segment with shorter white hair. Head shining, smooth; the upper part covered with fuscous hair, which is much longer behind the ocelli; the lower part is more thickly covered with short silvery pubescence. Mandibles broadly yellow behind the middle, the two basal teeth piceous, the apical tooth darker in colour. The inner orbits with a yellow line; the clypeus yellow, the extreme apex piceous; the central tooth larger, the lateral not half its size; the palpi yellow. Thorax black, except two yellow marks on the scutellum; rather thickly covered with longish fuscous hair; the hair on the pleuræ and apex of the median segment white. Median segment transversely irregularly striated; in its centre is a somewhat triangular area which reaches near to the apex of the top part, where it is rounded, and from which a straight keel runs to the apex; down the centre of the triangular area are two keels not reaching the end of the area, and bulging out at the apex. At the top of the oblique apex are, on each side of the central keel, three areæ, the inner being the larger; the rest of the areæ irregular. Wings clear hyaline, the stigma and nervures black; the radial, the base of the first and the second and third cubital cellules, smoky. Legs thickly covered with short white hair; the anterior knees, tibiæ, and tarsi yellowish in front; the apices of the four hinder coxæ and of the trochanters, a line below the apex of the hinder tibiæ, and the spurs yellow; the femoral tooth oblique, stout, twice as long as broad, the apex bluntly rounded. Abdomen shining, impunctate, the apex and ventral surface sparsely covered with long black hairs; the spots on the second segment obscure yellow (perhaps discoloured); the third segment obscure testaceous laterally at the base.

GASTROSERICUS BINGHAMI, *sp. nov.* (Pl. 16, f. 8).

Long. 5 mm. ♂.

Hab. Barrackpore *(Rothney).*

This species differs in too many points from *G. Rothneyi* to be its ♂. It differs also from the type of the genus *(G. Waltii* from Egypt) in having the clypeus toothed in the middle.

Head alutaceous, the vertex covered with a pale golden microscopic down; the orbits behind with silvery pubescence; the face and oral region thickly covered with golden pubescence; the space where the hinder ocellus should be, shining, smooth. Clypeus yellow, the apex piceous; its centre raised and projecting into a stout triangular tooth; the mandibles yellow, piceous at the apex. Antennæ stout; the basal joints with a minute silvery pubescence; the apex of the scape yellow; the ocellar space raised and surrounded, except in front, by a furrow, and a wider furrow runs down the vertex. Thorax alutaceous; the scutellum finely punctured; the median segment at the apex finely punctured; the extreme apex minutely transversely striated; the fovea is wide and deep, sharply narrowed at the apex. The sides, base, and apex of the mesonotum thickly covered with golden hair; the hair on the meso- and meta-pleuræ silvery. Tegulæ and a curved spot at the apex of the pronotum yellow. Wings hyaline, the nervures and costa fuscous, darker towards the apex. Legs black; the apical half of the fore femora, the apices of the four hinder and the tibiæ and tarsi clear yellow; the four hinder tibiæ broadly lined with black behind at the base; the 3—5 joints of the hinder tarsi infuscated. Abdomen covered with a sericeous pubescence; the sides of the dorsal and the apices of the apical ventral segments obscure testaceous; the apical segment for the greater part rufous.

PISON ORIENTALE, *sp. nov.*

Long. 8 mm.

Hab. Barrackpore (*Rothney*).

Comes near to *P. striolatum*, but differs in the striolated metapleuræ. Entirely black; head and thorax thickly covered with long fuscous hairs; the face more closely covered with silvery hair. Front rugosely punctured; behind the ocelli the punctures much finer and more widely separated. Apex of clypeus shining, bare, the apex in the middle produced into a small rounded point. Antennæ covered with a white microscopic pile. Thorax thickly covered with longish whitish hair. Mesonotum bearing large distinctly separated punctures which are much closer together; scutellum with the punctures smaller and not so deep, and almost absent in the centre. Median segment with a wide furrow in the centre in which are a few stout transverse keels; on either side of this it is irregularly punctured and has some curved striæ. Propleuræ shining, strongly depressed obliquely in the centre; mesopleuræ strongly punctured all over, and without a distinct longitudinal furrow; metapleuræ almost impunctate and more shining than the mesopleuræ. Wings hyaline, the costa and stigma black, the other nervures not so deep in tint; the recurrent nervures received shortly in front of the transverse cubital. Abdomen thickly covered with white hair, which is especially thick at the sides of the segments at their apices; the basal segments sparsely punctured; the others impunctate; the basal ventral segment strongly punctured.

PISON APPENDICULATUM, *sp. nov.* (Pl. 16, f. 5).

Long. 7—8 mm.

Hab. Barrackpore (*Rothney*).

Resembles *P. orientale*, but has the body more thickly pilose, the apex of the clypeus more broadly produced in the middle, the appendicle of the areolet as long as the cellule itself, and the recurrent nervures are received at a greater distance from the transverse cubitals.

Head closely and rather strongly punctured, more closely and hardly so strongly behind the ocelli; the front and vertex covered with long fuscous hair; from the lower part of the eye incision to the apex of the clypeus thickly covered with longish silvery hair, which hides the sculpture entirely; the apex of the clypeus roundly produced in the middle. Thorax thickly covered with long fuscous hair; the mesonotum strongly punctured, particularly at the sides; the centre of the scutellum almost without punctures; the median segment at the base shagreened, the centre with a wide, deep furrow, in which are a few stout transverse keels; the apex is irregularly and rather strongly transversely striated, the striations coarser above than below; there is a deep furrow at the top. Propleuræ coarsely shagreened; the mesopleuræ strongly punctured and without a furrow; the basal half of the metapleuræ more shining and less pilose than the rest. Wings hyaline, the nervures and stigma blackish; the pedicle of the petiole oblique, as long as the cellule; the recurrent nervures are received somewhat less than half the length of the cellule in front of the transverse cubital nervures. Legs thickly covered with longish hair and white pile; the spurs pale testaceous at the base. Abdomen thickly haired, and, at the apices of the segments, lined with silvery pubescence.

PISON (PARAPISON) CRASSICORNE, *sp. nov.*

Long. 5 mm. ♀.

Hab. Barrackpore (*Rothney*).

Comes near to *P. Rothneyi*, but smaller, the furrow on the median segment extending from the base to the apex; and the apex transversely striolated.

Head: the vertex and front shagreened, covered with a short fuscous pubescence; the cheeks and clypeus thickly covered with silvery pubescence; the mandibles and palpi pale testaceous. Antennæ entirely black, distinctly thickened towards the apex. Thorax shining, the pro- and meso-thorax impunctate, the oblique half of the median segment transversely striated; there is at its base a narrowish furrow, and behind the striated part, and separated from the basal, is a short, wider, and deeper furrow. The pro- and meso-pleuræ are shagreened; the latter has a wide and deep longitudinal furrow in the middle; the metapleuræ are shagreened at the base; the rest smooth and shining. The second cubital cellule at the top is as wide as the space bounded by the recurrent and first transverse cubital nervures; the upper part of the second transverse cubital nervure is curved; the lower straight, only slightly oblique; the second recurrent nervure is interstitial. The four front tibiæ are for the greater part dark testaceous; the hinder pair broadly dark testaceous at the base; the calcaria pale. Abdomen shining, impunctate, densely covered with white pubescence towards the apex; the five apical segments cream-coloured at their apices, the last more broadly than the others.

Trypoxylon cognatum, sp. nov.

Nigrum, abdomine rufo-balteato, capite thoraceque dense albo-pilosis, calcaribus albis; alis hyalinis, apice fere fumato.
♀. Long. 11 mm.

Hab. Himalaya.

Head black; the front and vertex alutaceous; the former with a shallow longitudinal depression in the centre; the eye incisions and the clypeus and the space below the antennæ densely covered with silvery pubescence; the vertex covered with short fuscous pubescence; the outer orbits except at the top, covered with longish, silvery pubescence; the mandibles piceous towards the apex; the palpi pale yellow. Antennæ entirely black, the scape covered with white hair. Thorax black; the mesonotum very shining, and with a bluish tinge; the pubescence on the pro- and meso-notum and scutellum dense, pale, that on the post-scutellum longer than on the scutellum. At the base the median segment is longitudinally striated; in the centre depressed, and in the middle of the oblique part is a longer, wider, and deeper depression; the apex closely punctured. Pro- and mesopleuræ and sternum densely covered all over with dense white pubescence; the metapleuræ sparsely covered with shorter white hair at the apex. Legs entirely black, except the apices of the four anterior tarsi, which are rufous, and the spurs, which are white. Wings clear hyaline, the apex slightly infuscated; the stigma and costa black; the nervures paler. Abdomen densely covered with short, pale hair; the petiole longer than the second and third segments united; the apex of the petiole and the second and the third segments ferruginous.

Comes nearest to *T. rejector* Sm. from Mainpuri, with which it agrees in coloration, but which differs from our species in having "an impressed line in front of the anterior ocellus, terminating at an elevated carina just above the insertion of the antennæ" (*cf. Trans. Zool. Soc. vii., 189*).

Explanation of Figures in Plate 16.

1. *Alyson annulipes.*
2. *Didineis orientalis.*
3. *Diodontus striolatus.*
4. *Dolichusus clavipes.*
5. *Pison appendiculatum.*
6. *Ichneumon mitra.*
7. *Cryptus perpulcher.*
8. *Gastrosericus Binghami.*
9. *Chrysis furiosa.*
10. *Hedychrum lugubre.*

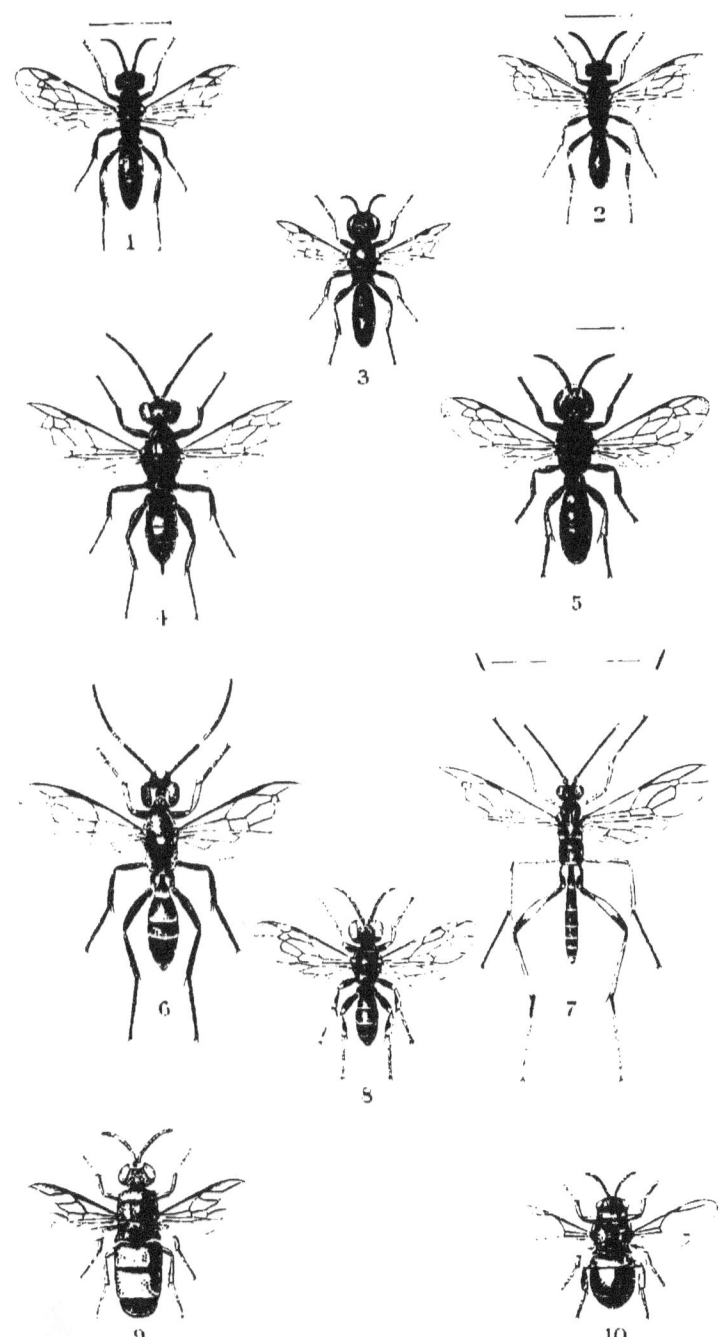

www.ingramcontent.com/pod-product-compliance
Lightning Source LLC
Chambersburg PA
CBHW051246300426
44114CB00011B/906